IUBS Section of Comparative Physiology and Biochemistry
1st International Congress, Liège, Belgium, August 27–31, 1984

Conference Organization

Organizing Board
R. Gilles, Chairman, Liège, Belgium
M. Gilles-Baillien and L. Bolis, Liège, Belgium/Messina, Italy

Host Society
European Society for Comparative Physiology
and Biochemistry

Under the Patronage of
The European Economic Community
The Fonds National de la Recherche Scientifique
The Ministère de l'Education Nationale et de la Culture Française
The Fondation Léon Fredericq
The University of Liège

The European Society for Comparative Physiology and Biochemistry
The American Society of Zoologists
The Canadian Society of Zoologists
The Japanese Society for General and Comparative Physiology

The Congress has been organized in relation with the 100th Anniversary of the School of Comparative Physiology and Biochemistry of the University of Liège.

The proceedings of the invited lectures to the different symposia of the congress have been gathered in five different volumes published by Springer-Verlag under the following titles:

Circulation, Respiration, and Metabolism
Current Comparative Approaches
Edited by R. Gilles (ISBN 3-540-15627-5)

Transport Processes, Iono- and Osmoregulation
Current Comparative Approaches
Edited by R. Gilles and M. Gilles-Baillien (ISBN 3-540-15628-3)

Neurobiology, Current Comparative Approaches
Edited by R. Gilles and J. Balthazart (ISBN 3-540-15480-9)

Respiratory Pigments in Animals, Relation Structure-Function
Edited by J. Lamy, J.-P. Truchot, and R. Gilles (ISBN 3-540-15629-1)

High Pressure Effects on Selected Biological Systems
Edited by A. J. R. Péqueux and R. Gilles (ISBN 3-540-15630-5)

High Pressure Effects on Selected Biological Systems

Edited by
A. J. R. Péqueux and R. Gilles

With 37 Figures

Springer-Verlag
Berlin Heidelberg New York Tokyo

Dr. André J. R. Péqueux (Scientific Editor)

Laboratory of Animal Physiology
Institute of Zoology
University of Liège, 22, Quai Van Beneden,
B-4020 Liège, Belgium

Professor Dr. Raymond Gilles (Coordinating Editor)

Laboratory of Animal Physiology
Institute of Zoology
University of Liège, 22, Quai Van Beneden,
B-4020 Liège, Belgium

ISBN 3-540-15630-5 Springer-Verlag Berlin Heidelberg New York Tokyo
ISBN 0-387-15630-5 Springer-Verlag New York Heidelberg Berlin Tokyo

Library of Congress Cataloging-in-Publication Data. Main entry under title: High pressure effects on selected biological systems. "Proceedings of the invited lectures to the First International Congress of Comparative Physiology and biochemistry organized at Liège, Belgium in August 1984"–Frwd. 1. Hydrostatic pressure–Physiological effect–Congresses. 2. Marine fauna–Physiology–Congresses. I. Péqueux, A. II. Gilles, R. III. International Congress of Comperative Physiology and Biochemistry (1st : 1984 : Liège, Belgium) QP82.2.P7H54 1985 574.19'135 85-22100

Printing and bookbinding: Beltz Offsetdruck, Hemsbach/Bergstr.
2131/3130-543210

FOREWORD

This volume is one of those published from the proceedings of the invited lectures to the First International Congress of Comparative Physiology and Biochemistry I organized at Liège (Belgium) in August 1984 under the auspices of the Section of Comparative Physiology and Biochemistry of the International Union of Biological Sciences. In a general foreword to these different volumes, it seems to me appropriate to consider briefly what may be the comparative approach.

Living organisms, beyond the diversity of their morphological forms, have evolved a widespread range of basic solutions to cope with the different problems, both organismal and environmental, with which they are faced. Soon after the turn of the century, some biologists realized that these solutions can be best comprehended in the framework of a comparative approach integrating results of physiological and biochemical studies done at the organismic, cellular and molecular levels. The development of this approach amongst both physiologists and biochemists remained, however, extremely slow until recently. Physiology and biochemistry have indeed long been mainly devoted to the service of medicine, finding scope enough for their activities in the study of a few species, particularly mammals. This has tended to keep many physiologists and biochemists from the comparative approach, which demands either the widest possible survey of animal forms or an integrated knowledge of the specific adaptive featuresoof the species considered. These particular characteristcs of the comparative approach have, on the other hand, been very attractive for biologists interested in the mechanisms of evolution and environmental adaptations. This diversity of requirements of the comparative approach, at the conceptual as well as at the technological level, can easily account for the fact that it emerged only slowly amongst the other new, more rapidly growing, disciplines of the biological sciences. Although a few pioneers have been working in the field since the beginning of the century, it only started effectively in the early 1960's. 1960 was the date of the organization of the periodical *Comparative*

Physiology and Biochemistry by Kerkut and Scheer and of the publication of the first volumes of the comprehensive treatise *Comparative Biochemistry* edited by Florkin and Mason. These publications can be considered as milestones in the evolution of the comparative approach. They have been followed by many others which have greatly contributed to giving the field the international status it deserved. Since the 1960's, the comparative approach has been maturing and developing more and more rapidly into the independent discipline it now is, widely recognized by the international communities of physiologists, biochemists and biologists. It is currently used as an effective tool of great help in the understanding of many research problems : biological as well as clinical, applied as well as fundamental.

The actual development of the field and the interest it arouses in a growing portion of the biological scientific community led some of us to consider the organization of an international structure, bringing together the major representative societies and groups around the world, which would aim at the general advancement and promotion of the comparative approach. This was done in 1979 with the incorporation, within the international Union of Biological Sciences, of a Section of Comparative Physiology and Biochemistry. The first International Congress of CPB, I organized in Liège with the help of a few friends and colleagues, is the first activity of this newly founded Section. In 22 symposia it gathered some 146 invited lectures given by internationally renowned scientists on all major current topics and trends in the field. The proceedings of these lectures have been collected in 5 volumes produced by Springer-Verlag, a publisher long associated with the development of CPB. The organization of the CPB Section of IUBS, its first Congress and these proceedings volumes can well be considered as milestones reflecting the international status and the maturity that the comparative approach has gained, as a recognized independent discipline, in the beginning of the 1980's, some 20 years after it was effectively launched.

Finally, I would like to consider that the selection of Liège for this first International Congress has not been simply coincidental. I thus feel that this brief foreword would not be complete without noting the privileged role Liège has played in some events associated with the development of the comparative approach. Liège had a pioneer in comparative physiology already at the end of the last century with Léon Fredericq. With Marcel Florkin, Liège had its first Professor of biochemisty, one of the founding fathers of comparative biochemistry. These two major figureheads of the comparative approach founded and developed what is actually called the Liège School of Comparative Physiology and Biochemistry, which was, at the time of the Congress, celebrating its 100th anniversary. This school provided early support to the European Society for Comparative Physiology and Biochemistry organized by Marcel Florkin and myself some years ago. The society, still headquartered in Liège, was, with the CPB division of the American Society of Zoologists, at the origin of the formation of the CPB Section of IUBS under the auspices of which this first International Congress, specifically devoted to the comparative approach, has been organized. An essential particularly of the Liège school of CPB is that its two founding fathers, scientists interested in general, basic aspects of the organization of living organisms, were also professors at the faculty of medicine. This largely contributed in Liège to avoiding the undesirable structuration of a so-called "zoophysiology" or "zoobiochemistry" independent of the rest of the field. The conditions were thus realized very early in Liège for CPB to play its key role in canalizing the necessary interactions between the general, pre-clinical or clinical and the environmental, ecological or evolutionary tendencies of physiology and biochemistry. The possibility of stimulating such interactions has served as a major guide line in the selection of the symposia and invited lectures from which these proceedings have issued.

R. GILLES
August, 1985

PREFACE

Of the various themes developed at the First International Congress of Comparative Physiology and Biochemistry, one dealt with some aspects of adaptation to environmental factors, one session being particularly concerned with the effects of hydrostatic pressure on selected biological processes. This volume is a collection of proceedings of the invited lectures of this session.

Why a study of the *influence of pressure on biological mechanisms ?*

In the natural aquatic environment, the hydrostatic pressure to which an organism is subjected increases by about 1 atm for every 10 m depth of water. In the greatest depth of the ocean, hydrostatic pressure thus ranges at about 1000 atm. If such extreme conditions have to be regarded as exceptional in the natural environment, 400 atm is, on the contrary, an order of magnitude frequently encountered, as about 50% of our plant lies under the ocean's mean depth, 4000 m.

All aquatic organisms have thus to face hydrostatic pressures much higher than those at which most of the experimental scientific research has been done. When one considers the fact that, according to many experimental and theoretical studies, pressure has to be regarded as a physical parameter almost as important as temperature, one is surprised that its importance as an environmental variable has not always been and is not yet actually widely recognized by all biologists.

In addition, special problems due to pressure arise in surface diving organisms that contain gas-filled cavities, including man, for whom they are of particular interest.

Beyond the idea of investigating the biological effects of high pressure from the ecological or ecophysiological point of view, some biologists have also used pressure as a simple experimental condition, as a tool to disturb selected biological processes in order to better understand the features of their structure and functioning at atmospheric pressure. It is, in fact, not always easy to clearly distinguish between all motivations.

The aim of this volume is certainly not to reflect all aspects of the biological effects of hydrostatic pressure. We intend to essentially restrict this review to the current knowledge of authors whose chief concern has been to understand the action of hydrostatic pressure on bio-membranes and related processes. In consideration of the basic importance of membrane phenomena in almost all physiological events, it is clear that the knowledge of how hydrostatic pressure affects membrane processes is essential to a thorough understanding of the adaptations of cells to deep aquatic environment conditions, of the physiology of almost all aquatic organisms, and indeed, even of air-breathing vertebrates, including human beings.

The symposium centered on two main topics.
The *first* series of papers will be directly concerned with *the effects of hydrostatic pressure on osmoregulation and ion transport across cell membranes*. It will include discussions of various facets of these effects from the organismic level down to the cellular level, even to the molecular level.
The *second* part of this volume will consider *the effects of pressure on electro- and neurophysiological processes* related to membrane properties, with special attention to the symptoms of heightened neural excitation displayed by vertebrates subjected to high pressure, the so-called High Pressure Nervous Syndrome (HPNS).
This volume does not claim to be complete by presenting extensive reviews covering all aspects of the topics approached. Its basic aim is to convince the reader that membrane function at elevated pressure is a central problem, not only for marine biology, but also for other more particular aspects of underwater physiology, like diving physiology, whose ecophysiological significance is perhaps less evident.

It is hoped that this book will be of interest to scientists concerned with various kinds of barobiology problems and that it will act as a stimulus for new research in the high pressure field, taking cell membrane phenomena into consideration.

Liège, Belgium, André J.R. PÉQUEUX.
December 1985

CONTENTS

LIST OF CONTRIBUTORS

R.W.BRAUER	Institute of Marine Biomedical Research University of North Carolina at Wilmington 7205 Wrightsville Avenue Wilmington, North Carolina 28403 U.S.A.
J.A.DUTCHER	Institute of Marine Biomedical Research University of North Carolina at Wilmington 7205 Wrightsville Avenue Wilmington, North Carolina 28403 U.S.A.
D.J.HARRIS	ARE Physiological Laboratory Fort Road Gosport, Hampshire PO12 2DU U.K.
K.HEREMANS	Laboratory of Chemical and Biological Dynamics Katholieke Universiteit te Leuven B-3000 Leuven BELGIUM
P.M.HOGAN	Department of Physiology, School of Medicine, State University of New York at Buffalo 123 Sherman Hall Buffalo, New York 14214 U.S.A.
E.D.JOHNSON	Institute of Marine Biomedical Research University of North Carolina at Wilmington 7205 Wrightsville Avenue Wilmington, North Carolina 28403 U.S.A.
M.R.JORDAN	Institute of Marine Biomedical Research University of North Carolina at Wilmington 7205 Wrightsville Avenue Wilmington, North Carolina 28403 U.S.A.
J.J.KENDIG	Department of Anesthesia Stanford University School of Medicine Stanford University Medical Center Stanford, California 94305 U.S.A.

A.G. MACDONALD
Department of Physiology
Marischal College
Aberdeen University
Aberdeen AB9 1AS
SCOTLAND, U.K.

Ch.G.MILLER
Institute of Marine Biomedical Research
University of North Carolina at Wilmington
7205 Wrightsville Avenue
Wilmington, North Carolina 28403
U.S.A.

J.L.PARMENTIER
Anesthesia Research Laboratory
University of South Alabama
284 Cancer/Clinical Sciences Building
Mobile, Alabama 36688
U.S.A.

A.J.R.PEQUEUX
Laboratory of Animal Physiology
University of Liège
22, Quai Van Beneden
B-4020 Liège
BELGIUM

R.D.ROER
Institute of Marine Biomedical Research
University of North Carolina at Wilmington
7205 Wrightsville Avenue
Wilmington, North Carolina 28403
U.S.A.

M.E.SHEEHAN
Institute of Marine Biomedical Research
University of North Carolina at Wilmington
7205 Wrightsville Avenue
Wilmington, North Carolina 28403
U.S.A.

C.J.SHELTON
Department of Biochemistry
Royal Free Hospital Medical School
Rowland Hill street
London NW3 2PF
ENGLAND, U.K.

CONVERSION TABLE

psi_a	atm_b	Kg/cm^2	$N/m^2{}_c$	
14.696	1	1.033	1.01325	$\times 10^5$
1,000	68	70	68.9	$\times 10^5$
5,000	340	352	344.5	$\times 10^5$
8,000	544	562	551.2	$\times 10^5$
10,000	680	703	689.0	$\times 10^5$
15,000	1,021	1,055	1,034.5	$\times 10^5$

a : Pounds per square inch

b : Atmosphere

c : Newtons/meter^2

INTRODUCTORY SURVEY

INTERACTION OF TEMPERATURE AND PRESSURE IN INTACT ANIMALS

R. W. BRAUER, M.R. JORDAN, Ch.G. MILLER,
E.D. JOHNSON, J.A. DUTCHER, M.E. SHEEHAN.

I. INTRODUCTORY CONSIDERATIONS

Repeatedly over the past 150 years physiologists and biochemists have encountered situations in which agents the effects of which on simple chemical systems were presumed to be thoroughly understood and readily quantifiable in terms of prevailing physiochemical theory were found to produce profound and also quantifiable biological effects. Such situations have tended to encourage research programs predicated on the assumption that in view of the "simple" nature of the agent employed - or the parameter manipulated - a careful analysis of what might be termed stress/strain relations should have an excellent chance of describing and identifying the particular component or response of the organisms studied which must hold the key to the observed effects. In more modern terms, it was hoped that such research programs could yield in one relatively simple step significant insight into the molecular mechanisms controlling some particular regulatory mechanism. Time has tended to deal unkindly with such hopes : in case after case the hoped-for simplicity perceived at the outset has resolved itself into increasingly complex successive arrays of interacting factors progressively blocking attainment of the originally perceived goal in a welter of increasingly detailed explanations of ever more restricted scope. Despite these many frustrations, the process as a whole has been a powerful generator of experimental programs which, though often failing in their original intent, on the whole have contributed significantly to the development of biophysical knowledge.

It seems to me that the study of the effects of high hydrostatic pressures, and of certain pharmacological phenomena related thereto, provides a good example of such a situation. The studies of pressure effects on bioluminescence and on contractility of the early half of this century were clearly inspired by early work of pressure effects upon proteins and by the elaborate theoretical apparatus to which this gave rise. Similarly, the more recent series of studies of pressure effects on the brain and more particularly on the phenomenon of pressure reversal of anesthesia spawned - and in part were driven by - what now appear somewhat simplistic theories of the actions of these agents on hypothetical membrane systems. As in other cases, the several decades which have elapsed since those beginnings have left us with a legacy of biological data which reflect the complexity of the underlying molecular events, such as we have just heard them described (Heremans, 1985; Macdonald and Shelton, 1985). Not only were pressure effects properly so called revealed at much lower hydrostatic pressures than the early workers held possible, but pressure and anesthesia effects upon different (but still hypothetical !) sites were found to show widely different quantitative relations, often were non-linearly related and subject to previously unsuspected genetic influences, and the sum total of effects clearly includes some for which pressure/anesthesia antagonism fails to hold altogether (Cf.e.g. review by Brauer *et al.*, 1982).

As I have listened to successive papers on this subject, including those presented today, it seems to me that in every case, and at every level of organizational complexity nature has successfully beaten off the first onslought and managed to leave us far from any comprehensive understanding of any of the biological or even of the chemical phenomena. Yet such a review also cannot fail to impress us with the fact that much has been learned, and that in this respect high pressure research has proven itself a useful tool. While the early hopes of quick and simple solutions have long faded, the concept that imposition of environmental stress upon biological systems is a powerful dissecting tool often revealing unsuspected regulatory processes has, it seems to me, stood the test of time in this case, and provides a justification for pursuing this particular road of inquiry.

There is in addition a special attraction to this general approach in the case of hydrostatic pressure . We know of at least two manipulations each of which on its own right belongs in the general category of influences we have been discussing so far, but which are uniquely pertinent here because they are capable of reversing - or at least partially reversing - important pressure effects on biological systems. We have already alluded to one of these, the antagonism between high pressure and general anesthetics epitomized by the phenomena of pressure reversal of anesthesia, and of the protective effect of inert gas narcotics against high pressure convulsions. The second type of manipulation that ought to be capable of reversing many pressure effects on thermodynamic grounds is temperature displacement. In both cases, the observed or postulated antagonism should permit an experimental approach not unlike the familiar titration in chemical systems where the proportions of two reagents in a system are adjusted until a predetermined equilibrium is established or re-established. In the case of biological systems this approach has the advantage that it allows one to focus attention upon systems functioning near normal levels, and hence much less subject to the multiple buffering and regulatory efforts exhibited by a severely deranged organism. Since the IG/HP case has been discussed *in extenso* repeatedly in recent years, I would like here to focus attention upon the case of temperature pressure interaction.

A. Problem formulation

To begin this discussion I should like to briefly recapitulate some thermodynamic considerations which I believe can serve to indicate the direction our inquiry can take profitably, and the degree of resolution the sort of study I propose to discuss can be expected to yield. I would than like to briefly review the - to me rather disappointing - results of a frontal approach seeking direct measures of temperature/pressure interaction by conventional methods of quantitative physiology. Finally, I propose to report to you on a new approach using the titration concept based on behavioral indices, and the promising - if still at times puzzling - results this has produced over the last year or two.

B. Thermodynamic data

Temperature and pressure are the two dominant extensive thermodynamic parameters characterizing biological environments. Direction and magnitude of changes resulting from displacement of any stable physicochemical system are dominated by the free energy changes characteristic of the particular transition under observation, and hence by the equation $dF = VdP - SdT$ or its counterpart for $dF^{\neq}$ in the case of rate processes. Thus the molecular properties which will dominate the situation will be ΔV, or $\Delta V^{\neq}$, the volume change associated with the transition under study, for pure hydrostatic pressure effects, and ΔS, or $\Delta S^{\neq}$, the corresponding entropy changes for pure temperature effects. The familiar Clausius Clapeyron equation for pressure dependence of the melting point of a condensed phase can be rewritten in its most general form : $dT/dP = \Delta V_T/\Delta S_P$, and in this form is applicable to a wide range of biologically relevant properties and phenomena. To evaluate the quotient dT/dP from the data most commonly available it can be approximated by $\Delta T/\Delta P$ derived from a thought experiment in which a temperature displacement ΔT is followed by a pressure displacement ΔP, ΔP and ΔT being chosen small enough to avoid complications from non-linear or nonreversible change, and of such relative magnitude and sign that at the end of the experiment the property, or the composition of the equilibrium mixture under study, has been returned to its initial value. Calculations using this small displacement method to estimate dT/dP for a number of potentially relevant phenomena (cf. footnote[1]) show that ΔT/ΔP can vary from -2 to more than + 20°C/100 atm. The frequency distribution for the same values shows a peak around + 1°C/100 atm (Fig.1). While both range and peak of this distribution are to some extend functions of the arbitrarily chosen group of phenomena evaluated , there is reason to feel that if the list were expanded by adding other values of potential biological relevance the modal value would not be very different, and the overall range would be bound to be expanded even further.

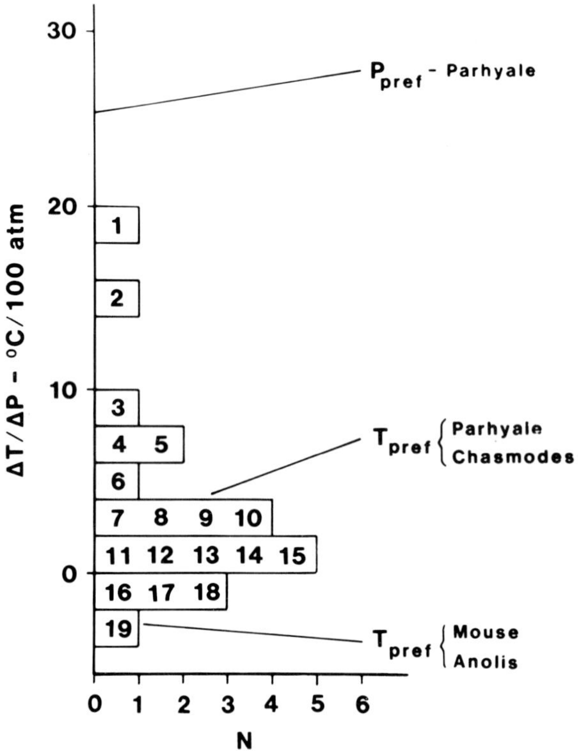

Fig.1. Frequency distribution of ΔT/ ΔP values for the transformations listed in footnote 1 together with estimates of corresponding ratios for the effect of pressure change on the T_{pref} for the four species listed and for the effect of temperature change on P_{pref} in Parhyale hawaiiensis.

From these figures one would predict that the overall effect of raising the pressure on a biological system by 100 atm should be not unlike that of a mere 1°C change in temperature, but that selected functions (the relative biological importance of which is open to conjecture) might show modifications which might reach magnitudes that in isobaric systems could be attained only by temperature shifts of 10° or 20°C. Clearly, there is no way to predict the magnitude of temperature/pressure interaction a biological system *a priori* on the basis of these data. Clearly, too, however , the potential range of values is such that if the magnitude of temperature pressure interactions with respect to some particular function could be measured, the result could contribute significantly to our understanding of the processes affected, the mechanisms of action, and the nature of the regulatory processes called into play.

II. DIRECT OBSERVATIONS OF T-P INTERACTION IN ANIMALS

A. Studies of temperature regulation in intact animals in high pressure atmospheres (cf.Sheehan and Brauer, 1979)

On the basis of earlier studies we concluded that thermal perception in high pressure environments is altered in the same sense, but to different degrees, in the mouse, the squirrel monkey, and man. The most complete data at 1 atm as well as at high pressure relate to the squirrel monkey and establish : 1. The euthermal interval is defined by the linear increase of metabolic rate with decreasing temperature below the lower critical point, and by a marked increase in heat flux through the skin of the foot as evidence of generalized vasodilatation when temperatures exceed the upper critical temperature. As monkeys are transferred from air to low pressure heliox environments, little change occurs in the width of this interval. As the pressure is increased from 5 to 30 atm, however, the width of the euthermal interval shrinks and remains small as pressures are increased further.

Within the euthermal interval the deep body temperature of the monkey decreases significantly (0.8°C), while hypothalamic temperature (T_h) remains at the control level, and mean skin temperature assumes a value which exceeds the ambient temperature by a constant small value (1.0°C). Hypothalamic responsiveness was measured by maintaining the animal at a temperature just above the level for vasodilatation and then cooling the hypothalamus, by an implanted thermode, until the first evidence of increased oxygen consumption was detected. Hypothalamic responsiveness, as indicated by the thermal displacement required for this effect, increases greatly as the pressure to which the animal is subjected is increased. The amount by which T_h decreases closely parallels the decrease in the width of the euthermal interval.

Taken together, the data indicate that in the euthermal interval temperature distribution within the animal is significantly different in high pressure heliox than at 1 atm. Furthermore, the close parallelism between hypothalamic responsiveness and the width of the euthermal interval are compatible with the hypothesis that the changes in the euthermal interval are directly attributable to changes in central thermal perception. This conclusion is further supported by the fact that in the interval over which hypothalamic responsiveness was tested, skin temperatures were the same at all pressures.

In addition to these changes, which relate to thermoregulation as such, changes in ambient pressure are associated with significant changes in oxygen consumption. In the mouse, as well as the monkey, these changes are characterized by two events : at low to moderate pressures, metabolic rate increases virtually linearly with pressure. This is true even if the temperature at which the metabolic rates are measured in each species is poised within the euthermal interval, i.e. at temperatures giving the lowest metabolic rates attainable at any given pressure. The magnitude of the increases in V_{O2} in man, monkey, and mouse appears to vary with basal metabolic rate, or inversely with body weight and similar parameters. At higher pressures, beginning (perhaps significantly) at the pressure at which a given species begins to show the high pressure neurological syndrome (HPNS) this relation between pressure and metabolic rate breaks down. The rate of oxygen consumption ceases to increase further as pressure is increased beyond this point and, at pressures approaching those giving rise to convulsions, actually shows a decrease.

Several comments are in order in regard to the V_{O2} data :

1. The increase in metabolic rate with pressure cannot be attributed secondarily to central activation of extrathermogenesis since it cannot be abrogated by increasing chamber temperatures to

well above the upper critical temperature, nor, while holding the animal within the euthermal ambient temperature range, by increasing hypothalamic temperatures until vasodilatation and symptoms of thermal distress are observed.

2. Data available for the mouse establish that this increase in oxygen uptake rate with increasing pressure cannot be attributed to increased respiratory work in the high density atmospheres. Comparing isonarcotic atmospheres of widely different densities, no differences in metabolic rate are observed over the range in which the correlation between pressure and metabolic rate is observed.

With regard to the lack of dependence of V_{O2} upon pressure in the high pressure interval, sufficient evidence is available to justify the conclusion that this is not attributable to limitations of respiratory capacity of the animals. The relevant facts include the following : (a) The high metabolic rates in this high pressure range can be sustained for many hours. Thus, fatigue does not seem to be a factor; (b) Studies with isonarcotic gas mixtures of differing densities in the mouse show that gas density becomes a factor limiting oxygen exchange only at gas densities well in excess of those which characterize the point of breakdown of the proportionality between pressure and oxygen consumption; (c) In the squirrel monkey even at high pressures, metabolic rates can be increased severalfold over what we are calling the resting level, either by cooling the chamber below the lower critical temperature, or by cooling the hypothalamus, or, finally , by pushing the pressures to the point at which the animal has a convulsion.

In reviewing this body of data, we infer that the increase in metabolic rate observed as a result of compression reflects some basic effect of pressure on metabolic rate rather than being secondary to any response to the environment as such. We feel that the data invite comparison with molecular mechanisms called upon to explain extrathermogenesis in cold acclimated animals or the extrametabolism elecited by thyroxin (Edelman, 1976).

These observations show that the energy requirements of sodium ion transport between the cells and the surrounding fluid are of such magnitude that they might also account for the changes. The hypothesis we find most appealing would be that a pressure induced increase in influx of some ion, such as sodium or chloride, elicits a compensatory increase in the corresponding transport ATP-ase activity to provide the basis for the increase in metabolic rate. A second component of such a model would hypothesize that the ion transport ATPase-involved, in its turn, suffers reduction in activity at pressures in the HPNS range, thus limiting the extra oxygen consumption attainable and at the same time accounting for a progressive distortion of the intracellular electrolyte make-up. Sodium ion is an appealing candidate for such a role since : (a) It has been shown to be involved in several instances of analogous mechanisms (Ismail-Beigi and Edelman, 1971); (b) the increase in intracellular sodium in neurons would provide the very type of electrolyte change that would make for increased excitability and epileptiform activity; (c) changes greatly resembling those postulated here have, in fact, been reported as the result of studies of the effect of pressure on the frog skin (Péqueux, 1976); (d) the type of changes that would result from the action of such a mechanism is compatible with the changes in thermoregulatory activity of the hypothalamus reported above, if one applies the results of studies manipulating sodium/calcium ratios in the tissue (Myers *et al.*, 1976).

B. Temperature effects on HPNS convulsion pressures

Another approach to the study of temperature/pressure interactions in intact animals makes use of the phenomenon of high pressure convulsions. In the squirrel monkey pressures of 40 or more atmospheres profoundly affect the CNS as shown by the tremors and convulsions of the high pressure neurologic syndrome. Changes in chamber temperature over the few degrees range which constitutes the euthermal interval at such pressures produce no recognizable change in convulsion threshold pressures. Temperatures outside this range produce such drastic behavioral and metabolic changes as to preclude effective study.

The mouse is a more tolerant subject : in baby mice a range of nearly 10°C can be explored. Yet, we have shown that in adult or baby mice over the entire temperature range which can be explored at high pressure without irreversible injury to the animals convulsion threshold pressures remain constant (Brauer *et al.*,1977; Mansfield and Brauer, 1980).

Even wider temperature ranges are accessible for study in truly ectotherm vertebrates. Among these, the lizard *Anolis carolinensis* shows no temperature dependence of convulsion threshold pressures. In two aquatic vertebrates, a newt and a fish, HPNS convulsion threshold pressures were found to be independent of test temperature, but to vary with acclimation temperature though the pattern of the relation varied profoundly between the two species (Beaver and Brauer, 1981) (Fig.2). Thus, all the vertebrate data so far are in accord in failing to reveal any direct effect of change of test temperature on high pressure convulsion thresholds over a temperature range of at least 10°C.

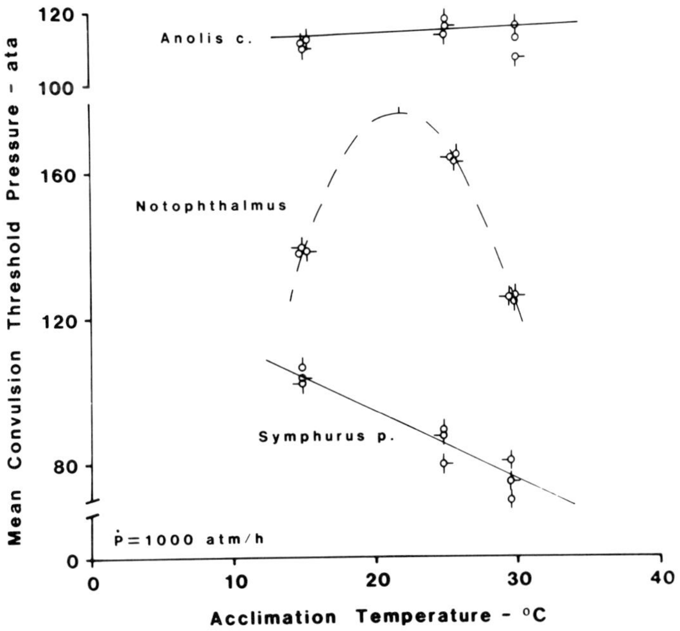

Fig.2. *Changes of HPNS convulsion threshold pressures with acclimation temperature in three ectotherm vertebrates. Tests were carried out in each case at temperature 5°C below (♀), at (O-), and 5°C above (O) acclimation temperature.*

C. Acclimation to high pressure.

1. Acclimation vs adaptation in deep water fauna

Does temperature/pressure reciprocity extend to long term effects so that, for instance one could demonstrate pressure acclimation ? That some sort of pressure accommodation must exist follows directly from the existence of a deep sea fauna, at pressures up to 1000 ata, as well as from various observations showing that pressure tolerance varies with habitat depth in a number of fish and crustaceans (Brauer *et al.*, 1980, 1984; Macdonald and Gilchrist, 1982). While in general such observations do not allow one to distinguish acclimation from adaptation effects, in the special case of certain aberrant Baikalian cottoids which have changed their habitat and yet retain at least in part the pressure tolerance characteristics of their "normal" relatives (e.g. species of normally abyssal cottoids inhabiting shallow waters), some assessment of the relative importance of the two factors is possible (Brauer *et al.*, 1984). In the one case for which we have adequate data about 60% of the observed pressure tolerance seems to be accounted for by past adaptation carried over into the new depth environment.

2. Laboratory studies of high pressure acclimation

More direct evidence concerning pressure acclimation comes from laboratory experiments involving multi-week exposure of animals to high pressures. For crustaceans acclimation with respect to pressure effects on osmoregulation appears rapid and extensive. The effects of such acclimation do not, however, seem to extend to growth impairment (or disturbance of moulting) by pressures of 100 to 150 ata (cf. Roer - this book).

Mice exposed to high pressure heliox for up to 2 weeks show rapid increases in mean convulsion threshold pressure : acclimation to 50 ata raises this from a control value of 93 to 115 ata; for 80 ata acclimated mice the corresponding value is 133 ata (Fig.3) (Brauer *et al.*, 1984). These changes are paralleled by equally marked decreases in tolerance to N_2O or N_2 narcosis. The range and scope of such pressure acclimation is limited : even after acclimation CD-1 mice cannot be maintained for more than a day or two at 100 ata in heliox, nor is there remission of high pressure anorexia. The effect of pressure acclimation is dissipated rapidly after the animals have been decompressed to 1 ata; within 24 hours convulsion threshold pressures have returned to values only slighly greater than those of control mice (Fig.3).

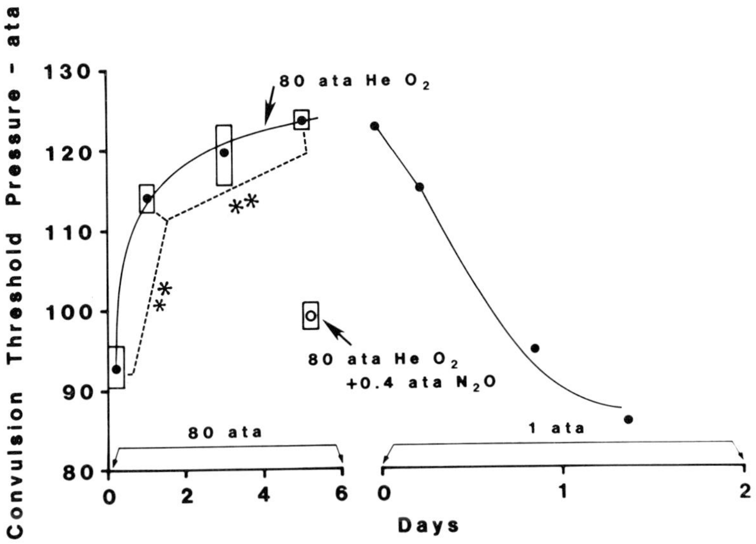

Fig.3. Time course of pressure acclimation of CD-1 mice to 80 ata heliox (●) and to 80 ata heliox containing 0.4 atm N_2O (O), and deacclimation on sojourn in air at 1 ata after 5 days at 80 ata in heliox.

Addition of as little as 0.5 atm of N_2O (or of 15 ata of N_2) to the heliox atmosphere largely prevents the increase in pressure tolerance resulting from prolonged exposure to 80 ata (Fig.3), as well as the associated decrease in anesthesia resistance proving that the effects are not related to gas density but to pressure effects on the CNS showing the familiar inert gas/high pressure antagonism.

Thus, high pressure acclimation is demonstrable but would seem to differ from temperature acclimation in being established and dissipated rather more rapidly. It is limited in scope to little more than 100 atm, and incomplete as shown by impaired growth in crustaceans and by persistence of high pressure anorexia in mice. Whether some of these characteristics will persist with longer conditioning exposures matching those needed for low temperature acclimation, remains to be shown.

D. Summary and implications

The data reviewed so far surely suggest that temperature regulation in mammals is substantially altered in high pressure atmospheres. However, the relative contributions to this from direct effects on the thermoregulatory centers, from the effects of high volume heat capacity in the dense atmospheres, and from more general CNS effects of high pressure underlying the high pressure neurologic syndrome is likely to prove extremely difficult to disentangle. The lack of effect of test temperature on convulsion pressures, and the complex acclimation and adaptation phenomena serve as further warnings that the systems being observed are highly buffered, and that temperature or pressure stress of such magnitude as to significantly displace the animals from a normal or comfort state are likely to becloud any analysis by multiple compensatory reactions which do not relate directly to the temperature or pressure effects one seeks to study.

Further analysis thus would seem to require more sensitive and less invasive methods, and might benefit greatly from the"titration" approach, seeking to manipulate pressure and temperature simultaneously in such a way as to maintain the animal in as nearly a "normal" state as may be achieved. We feel that behavioral methods hold great promise in this context and will now turn to the description of methods and results worked out to-date in this field.

III. USE OF BEHAVIORAL METHODS IN THE STUDY OF T-P INTERACTION

A. General description of T_{pref} methods

The primary tool we have used here has been exploration of the phenomena of behavioral temperature regulation and in particular determination of T_{pref}, the temperature preferendum. From among the various possible paradigms available for the determination of this parameter in various species we have chosen the simplest : the thermal gradient tube is essentially a straight tube in which an animal can move freely and along the axis of which the temperature changes at a constant rate to produce a stable temperature gradient of constant steepness (cf. Kinney *et al.*, 1981 and Brauer *et al.*,1984, for description of the design of some of the gradient systems we have employed). The position - and consequently the temperature - occupied by the test animal in such a tube is recorded at regular time intervals chosen small enough to assure that the motor behavior is adequately described by the sequential positions. The resulting body of data is used to construct for each experiment a cumulative frequency plot as a function of ascending temperature, and this in turn allows determination of the temperature corresponding to the maximum slope of this integrated frequency curve, and hence of T_{pref}, the temperature of the peak sojourn frequency; from the slope at this point one can estimate the precision with which T_{pref} in that experiment is determined using the equation $1/\sigma_x = \sqrt{\pi}.d\phi/dx$. Finally, from the sequential position data one can compute net displacement for the time intervals between successive data points, and from these results construct a new curve showing mean displacement speed for the several temperature intervals. These latter curves typically show a pronounced though broad minimum and this is invariably centered in the temperature interval which includes T_{pref} (cf. Fig.4). Thus we can redefine the temperature preferendum as the temperature corresponding to both the maximum sojourn density and the minimum locomotor activity, of an animal placed in a linear temperature gradient tube.

B. Supplementary data regarding T_{pref}

1. Learning aspects

Efficient display of the basic behavior pattern requires a certain amount of learning. We have chosen the linear gradient systems in preference to operant conditioning devices precisely because this paradigm calls for responses closest to those utilized by the animals in their natural setting and should therefore require a minimum of training and be applicable to the widest possible range of species. Both of these hopes have been born out. Still, naive mice first placed into such a system require several hours before discovering that they are indeed in a temperature gradient when they abruptly alter their behavior to achieve maximum comfort. The time required for this is sharply reduced with repeated exposures. Learning effects have been observed in reptiles as well as in fish so that as a general practice we use animals trained to each gradient system.

2. Comparative data

Table 1 shows T_{pref} for a number of species we have explored under habitat conditions. (see further in the text).

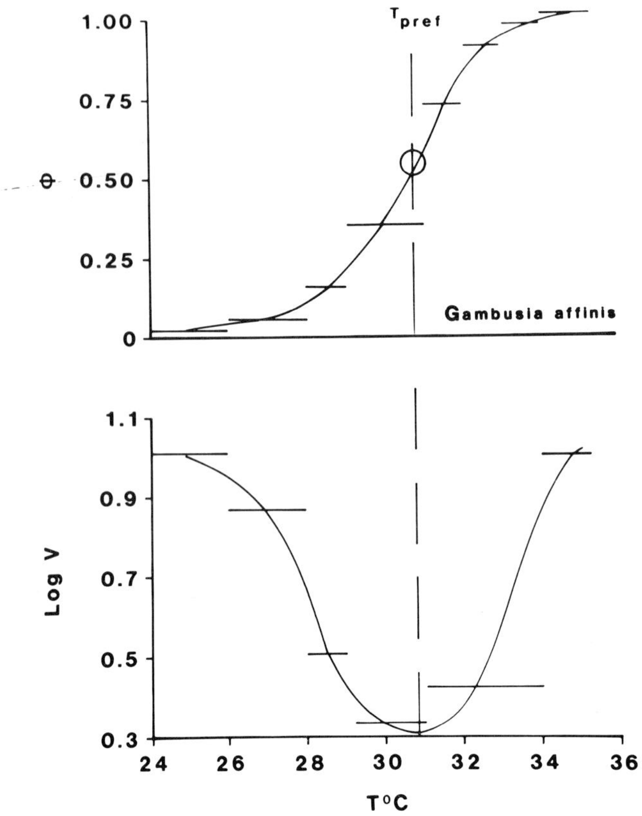

Fig.4. *Temperature preference behavior of Gambusia affinis. Upper panelcumulative frequency diagram. Lower panel - mean speed of displacement corresponding to the gradient segments shown (plotted as log V; units of V-cm/min).*

3. Hypoxia and T_{pref}

In all species so far tested we have encountered a common pattern of dependence of T_{pref} on PO_2 in the surrounding medium. In mice, lizard, fish, and gammarid crustaceans alike T_{pref} is independent of PO_2 between 0.1 and 0.2 atm. When oxygen partial pressures are reduced further, however, the temperature preferenda decrease sharply with further deepening of the hypoxia in all of these species (Fig.5). This decrease in the preferred temperature can be reversed almost instantaneously by replacing the hypoxic medium with a normoxic one. Our present interpretation of this effect of hypoxia in the mouse,

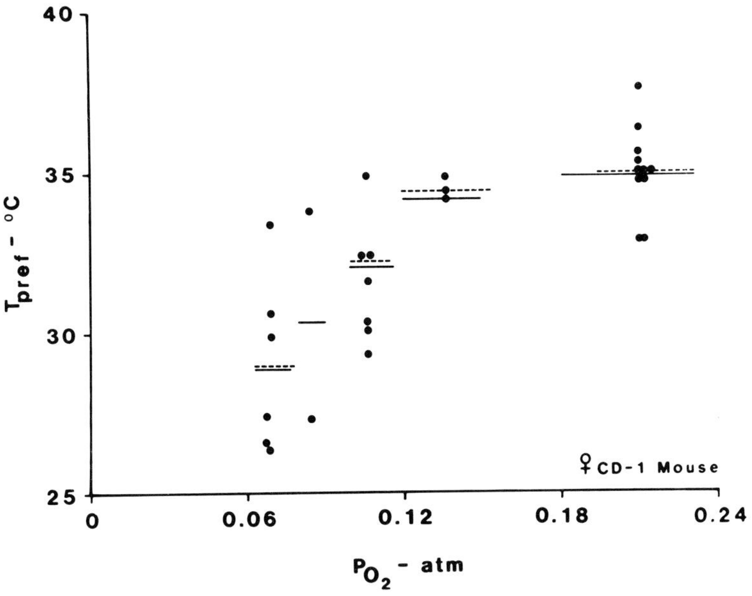

Fig. 5. *Preferred temperature of ♀ CD-1 mice as a function of oxygen partial pressure. --- Mean value and individual values determined at 1 ata in Heliox; — mean value determined at 80 ata in Heliox.*

based on a number of different lines of experimentation, is that it is exerted at the central nervous system level rather than upon peripheral thermogenesis or energy production. Furthermore, the fact that the effect is common to species as diverse as mouse and planarium suggests that at least one component of the temperature sensing system involved in temperature preference behavior over a large part of the animal kingdom must be strongly dependent upon aerobic reactions. For our present purposes the principal relevance of these observations is confined to helping to define suitable test conditions for studies concerning pressure effects on temperature preference behavior.

C. Pressure effects on temperature preferenda

1. *Parhyale hawaiiensis*

Parhyale hawaiiensis, is a small subtidal benthic gammarid amphipod with well defined temperature preference behavior (Kinney *et al.*, 1981). In this species increase of hydrostatic pressure inevitably entails an increase in the preferred temperature. To a reasonable degree of approximation the magnitude of this shift in T_{pref} is proportional to the pressure applied, the least square regression line having a slope of 4.2°C/100 ata and a correlation coefficient of 0.83.

During our second expedition to Lake Baikal these findings were extended to the giant abyssal gammarids found there at depths of 600-1400 m and hence accommodated to pressures averaging 100 ata (Brauer *et al.*, 1984). They thus provided an opportunity to study decreased as well as increased hydrostatic pressures. T_{pref} for these deep water gammarids was much lower than for *Parhyale* -- at habitat pressure it coincided with the temperature range of their habitat (3-4°C) -- but for pressure displacements in either direction from the habitat pressure, the preferred temperature shifted in the same direction and at the same rate as for shallow water forms, i.e. by +4°C per 100 ata pressure increase.

These data prove that animals which never in their lives had been exposed to recognizable temperature fluctuations, still retain temperature preference behavior which is fully as well defined as that of their shallow water cousins for whom wide temperature excursions are a daily experience. The fact that the same pattern of temperature/pressure interaction with respect to temperature preference was seen in the marine gammarid tested in seawater and in the Baikalian gammarids which come close to living in distilled water rules out osmoregulation as an important factor controlling temperature preference behavior in this genus. Finally, the slope of the T/P line is suggestive : it falls in the immediate vicinity of the value of T/P for variation with pressure of the melting point of lipid bilayer membranes as well as for pressure displacement of the breakpoint in the Arrhenius diagram for membrane-bound Na-K-ATP-ase from synaptosomes (Fig.1). This raises the possibility that in the case of these gammarids pressure dependence of temperature preference behavior may be related fairly directly to membrane properties of excitable cells.

2. CD-1 mouse

A very different set of data is available for the CD-1 mouse. Temperature preference behavior in this species is very well defined (cf.table 1). In high pressure heliox deep body temperatures generally follow chamber temperature and are 1.5 - 2.0°C above ambient as defined by an integrating thermometer ("copper mouse") (Sheehan and Brauer, 1975).

TABLE 1. Preferred temperatures at habitat pressure for selected species

Species	Test Pressure	Test Medium	$T_{pref} \pm \sigma$ °C
Crustacea			
Parhyale hawaiiensis	1 ata	seawater	20.1±1.5
Abyssal gammarid species (Lake Baikal) **	100 ata	lakewater	3.6±0.4
Vertebrata			
Pisces			
Gambusia affinis	1 ata	pondwater	29.9±1.8
Chasmodes bosquianus	1 ata	seawater	24.5±1.8
Reptilia			
Anolis carolinensis	1 ata	H_eO_2	34.5±0.7
Mammalia			
Microtus piretorum	1 ata	H_eO_2	32.7±0.5 *
Mus musculus CD-1	1 ata	H_eO_2	35.1±0.9 *
Mus rattus CD	1 ata	H_eO_2	32.7±0.5 *
Mesocricetus auratus	1 ata	H_eO_2	32.3±0.6 *

* These values of σ are intraindividual values computed from the slopes of the frequency curves. The remaining σ represent estimates of interindividual variability.

** Acclimation temperature 3.0-4.0°C. Acclimation temperatures for all remaining species are 22-24°C.

On compression, unlike the gammarids, mice select slightly reduced temperatures: in each of the eight animals the mean T_{pref} at 50 ata in heliox was lower than at 1 ata, the average difference being 1.4°C, significant at the 0.1% level. The data now available establish that in the mouse compression in heliox entails a reduction in T_{pref}, and that $\Delta T/\Delta P$ is about 2.8°C per 100 atm. Combining this with data on deep rectal temperatures at high pressure we find that at their preferred temperature our mice at 1 ata have T_R = 37.5 to 38.0°C while at 50 - 80 ata the T_R corresponding to T_{pref} is 36.5 - 37°C. This drop is statistically significant, and lends credence to the observation quoted above that in the eurythermal squirrel monkey deep body temperatures are about 0.5°C lower at 40 - 50 ata than at 1 ata. The hypoxia response of T_{pref} appears to be unaffected by compression in heliox to 50 ata (Fig.5) despite the fact that metabolic rates are at least three times greater than at 1 ata suggesting that P_aO_2 rather than some factor related to whole body V_{O_2} is the determinant for this effect.

3. *Anolis carolinensis* and *Chasmodes bosquianus*

Less extensive data are available for two additional ectotherm vertebrate species. *Anolis carolinensis* is a small (2-5g) lizard, effectively in thermal equilibrium with its environment, with a pressure tolerance slightly greater than that of the mouse (Fig.2). Animals acclimated to 22-24°C show well defined temperature preference behavior (Table 1). On compression to 80 ata *Anolis* show a 1.5°C decrease in preferred temperature. By extrapolation this corresponds to $\Delta T/\Delta P$ = -1.9°C per 100 atm and thus is similar to the mouse.

The swimbladderless blenny *Chasmodes bosquianus* is of appropriate size (small adult speciments are only 15 millimeters in length) to be satisfactorily accommodated in our small hydrostatically driven gradient tube systems and shows satisfactory temperature preference behavior (Table 1). The animals have a pressure tolerance slightly below 100 ata. Compression to 50 ata entails an unmistakable increase in the temperature preferendum averaging 1.8°C , extrapolating to +3.6°C per 100 atm, close to the value found for the other aquatic species, *Paryhale*, and unlike the terrestrial ectotherm vertebrate, the reptile.

D. Summary of T_{pref} Results

The observations available to-date concerning pressure effects on temperature preference behavior thus reveal a striking distinction in the temperature-pressure interactions which divide the four species studied to-date into two classes (Fig.1). At this particular point one can equally well describe the resulting grouping by focusing on the fact that the two aquatic species show high positive displacements of temperature preferenda with increasing pressure, while the two gas breathing species show somewhat smaller displacements in the opposite direction; or by noting that the one endotherm species tested and the one ectotherm species closest to it on the phylogenetic scale show thermoregulatory behavior characterized on compression by decreases of modest size of the preferred temperature, while the two species lower on the phylogenetic ladder show increases in T_{pref} on compression and thus appear more closely linked to recognizable biophysical changes such as might be presumed to affect excitable membranes. Further experiments, perhaps focusing first of all on amphibia, and second exploiting the indications of a metabolic link to temperature perception should help to clarify the position and prepare the ground for more analytical studies that may be able to exploit these differences in temperature/pressure linkage to shed new light on the complex mechanism giving rise to temperature perception and its behavioral expression.

E. Pressure Preference Behavior

Before closing I should like to tell you very briefly about yet another temperature/pressure interaction phenomenon we have only just begun to explore. To a significant extent the progress of experiments reported above was guided by our preoccupation with the reciprocity between temperature and pressure predicted on thermodynamic grounds. Since we have explored effects of pressure change on temperature preference behavior, it would thus seem natural for us to become curious about the possibility of studying temperature effects on pressure preference. In the past such studies seemed altogether impracticable. Given the limited rate of locomotion of most aquatic animals, naturally existing pressure gradients are far too shallow to permit meaningful studies. Production of the high gravity environments required to remedy this problem in the laboratory without incurring prohibitive levels of Coriolis forces would require centrifuges of truly awe-inspiring dimensions. Recently, however, with the development of fast-responding pressure amplifiers it occurred to us that, while it might be impracticable to create real pressure gradients in the laboratory, one could simulate such pressure gradients in a way which would leave one's test animals substantially unable to distinguish between simulation and reality : it would suffice to equip a tube similar to those used in the temperature gradient experiments with an appropriate computer-controlled pressure amplifier system programmed to regulate the pressure in the tube so that it would instantaneously translate any displacement of the animal into a change of pressure. The most useful coupling function would be such that an animal would experience a steady increase in pressure as it swam in one direction and a fall at the same rate as it swam in the opposite direction. At present by far the best computer we have found for achieving the required coupling between animal position and pressure in such systems is a well motivated graduate student. A number of successful experiments have been carried out with such a hybrid prototype device, using *Parhyale hawaiiensis*, our favorite pilot animal, in a simulated gradient of 7.5 atm/cm (Brauer *et al.*, 1984). These experiments have established that pressure preference behavior does indeed exist : the sojourn frequency distribution of these animals along a gradient tube changes from random in the absence of a pressure gradient to a well defined maximum frequency at some intermediate pressure when the gradient is applied, provided the pressure range has been chosen properly. If the gradient is moved along the tube, the animals follow it so that the maximum of the sojourn frequency distribution continues to fall at the same pressure as before. If the gradient tube is lengthened threefold, reducing gradient steepness to 1/3, the animals are still capable of orienting to the simulated pressure gradient, and still show maximum sojourn frequency at the same pressure. This pressure thus would appear to be properly designated as the pressure preferendum, P_{pref}. For *Parhyale* acclimated to 1 ata at 22-24°C the value of P_{pref} is quite reproducible from one animal to the next. It equals not 1.0 but rather 60±2 ata ! P_{pref} varies with temperature; differences between the results at 20° and at 30°C are well secured statistically ($P < 0.01$). However, with a mean coupling factor $\Delta P/\Delta T = +\ 2.8$ atm/°C the coupling between T and P in this case is much less close than for temperature preference behavior. Inverting and converting to the same units as were used above for T_{pref} we find $\Delta T/\Delta P = 36$°C/100 ata, at the upper extreme of the thermodynamic scale of Fig.1.

F. General summary of results of behavioral studies of T-P interaction

The completed Fig.1 sums up the results gained to-date regarding preferred temperature and preferred pressure: T_{pref} values fall into two groups, aquatic vs terrestrial, or fish and gammarid vs mouse and lizard, as you please, with $\Delta T/\Delta P$ values around +4.0 and - 2.0 °c/100 atm respectively. Recalling the thermodynamic data from which we started, Fig.1 shows that the "aquatic group" values correspond closely to characteristics of non-isotropic membrane systems that may be relevant for performance of excitable cells.

The data for the terrestrial animals, by contrast have no such suggestive molecular counterpart (the only comparable value in Table 1 refers to pressure activation of the ATPase of a deep sea fish) and may reflect superposition of peripheral input on a pressure sensitive integrative central reference mechanism involved in thermoperception. One may speculate that the same integrative system is responsible for the remarkably small effect of compression on deep body temperature in mammals. Pressure preference behavior has now been demonstrated for the first time and is startling against the background of T_{pref} data, both because of the high pressure consistently preferred by a shallow water gammarid, and because of the slight temperature dependence of P_{pref} which implies a value for $\Delta T/\Delta P$ well beyond the scale of relevant physicochemical phenomena we set up at the outset. Whatever the basis for this behavior may prove to be it is already clear that the biophysical basis for sensing absolute pressure must be radically different from that for temperature perception.

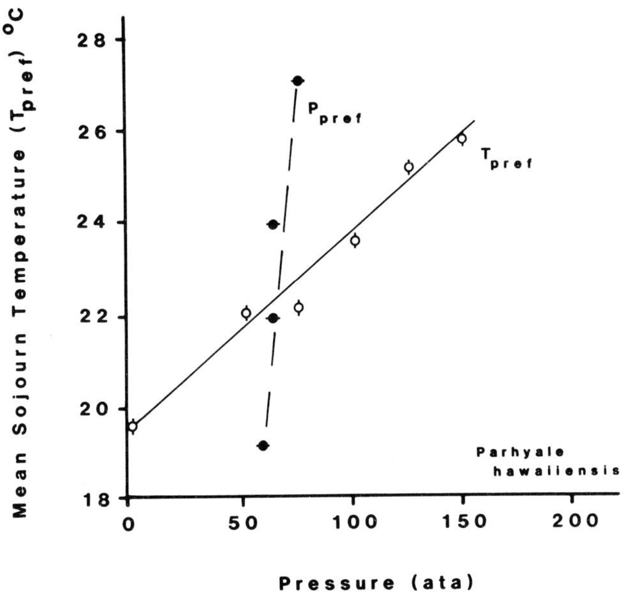

Fig. 6. *Temperature preferenda as a function of test pressure and pressure preferenda as a function of test temperature for the gammarid amphipod Parhyale hawaiiensis acclimated to 1 ata and 20-24°C.*

Footnote.

1) Estimated values of $\Delta T/\Delta P$ (°C/110 atm) for mutual neutralization of small temperature and pressure displacements in numerical order corresponding to Fig.1 are as follows : 1. Viscosity (Octane) : 20; 2. Density (H_2O) : 16; 3.Density (Octane) : 10; 4.Lethal temperature (Ciliates) : 8; 5.Arrest pressure (ciliary beat) : 8; 6.Tensile strength (cytogel) : 4.5; 7.Lipoprotein membrane melting point : 3; 8.Break in Arrhenius plot for Na/K-ATPase: 3; 9.Melting point (dodecane) : 2.5; 10. Specific activity Na/K-ATPase : 2.1; 11. Turtle Auricle (beat frequency): 1.0; 12.V_{O_2} (*Tetrahymena*) : 1.0; 13.Specific activity (isocitric dehydrogenase) (trout):0.8; 14. Sol-gel transformation (Me-cellulose) : 0.25; 15. Pressure Block of Cell Division (ciliates) :0.2; 16.Viscosity of H_2O : -0.1; 17. Dissociation of H_2O: -1.1; 18.Specific activity (Myosin ATPase) ; 1.9; 19.Specific activity (*Antimora* Na/K ATPase) (< 200 ata):-2.1.

Acknowledgement. This research has been supported by the Office of Naval Research through Contract N00014-75-C-0468, and the generous help of the Max and Victoria Dreyfys Foundation.

REFERENCES

Beaver, R.W., Brauer, R.W. (1981) Pressure/Temperature interactions in relation to development of high pressure convulsions in Ectotherm Vertebrates. Comp. Biochem. and Physiol. 69(A): 4, 665-674.

Brauer, R.W., Beaver, R.W., Lahser, S., Mansfield, W.M., Sheehan, M. E.(1977) Time, rate and temperature factors in the onset of high pressure convulsions. J. Appl. Physiol. Respir. Envir. Exercise Physiol. 43(2), 173-182.

Brauer, R.W., Jordan, M.R., Roer, R.D., Williams, E.E., Bekman, M.Y., Galasii, G.I., Sidelyova, V.G.(1984) Pressure effects on thermal preference behavior in gammarid amphipods from 600-1000m in Lake Baikal. Journal of Thermobiology ,(In press).

Brauer, R.W., Hinson, W.M., Dutcher, J.(1984) Modification of pressure and anesthesia tolerance by multi-day exposures to high pressure heliox or to sub-anesthesic levels of general anesthetics. In : Roth S. (ed) Molecular mechanisms of anesthesia III, Raven Press, New York . (In press)

Brauer, R.W., Hogan, P.M., Hugon, M., MacDonald, A.G., Miller, K.W. (1982) Patterns of interaction of the effects of the light metabolically inert gases with those of hydrostatic pressure as such - a review. Undersea Biomedical research, 9,. 353-395.

Brauer, R.W., Jordan, M.R., Kinney, M., Miller, G., Williams, E.E. (1984) Temperature/pressure interactions in regard to temperature and pressure preference behaviour. In : Hales JRS (ed) Thermal Physiology, Raven Press, New York, 499-502.

Brauer, R.W., Kaizer, D.J., Nesbitt, D.L., Rait, C.L., Bekman, M.Y. (1980) Reaktsia gammarid Baikala na Gidrostaticheskoe Davleniyie. Zh. Evolutsionnoi Biokhimii i Fiziologii 16, 545-549.

Brauer, R.W., Sidelyova, V.G., Dail, M.B., Galazii, G.I., Roer, R.D. (1984) Physiologic adaptation fo cottoid fishes of Lake Baikal to Abyssal Depths. Comp. Biochem. and Physiol. 77(A) 699-705.

Edelman, I.S. (1976) Transition from Poikilotherm to the Homeotherm : Possible role of sodium transport and thyroid hormone. Fed. Proc. 35, 2180-2183.

Heremans, K.(1985) Pressure effects on phospholipids and proteins. (This volume).

Kinney, M., Jones, W.R., Brauer, R.W., Sorrel, F.Y.(1981) A gradient tube system for the study of the effect of high hydrostatic pressures on temperature preference behavior in small aquatic animals. Comp. Biochem. and Physiol. 68(A), 501-505.

Ismail-Beigi, F., Edelman, I.S. (1971) The mechanism of the calorigenic action of thyroid hormone . J. Gen. Physiol. 57, 710-722.

Macdonald, A.G., Gilchrist,I. (1982) The pressure tolerance of deep sea amphipods collected at their ambient high pressure. Comp. Biochem. Physiol. 71A, 349-352.

Macdonald, A.G., Shelton, C.J.(1985) Ionic regulation under high pressure - some observations of comparative and theoretical interest. (This volume).

Mansfield, W., Brauer, R.W., Gillen, H.W., Nash, K. (1980) Changes in CNS response to high pressure during maturation of newborn mice. J. Appl. Physiol. Resp. Envir. and Exercise Physiol., 49(3), 390-397.

Myers, R.D., Gisolfi, C.V. , Mora, F.(1977) Role of brain Ca^{2+} in central control of body temperature during exercise in the Monkey. J. Appl. Physiol. 43, 689-694.

Péqueux, A. (1976). Polarization variations induced by high hydrostatic pressures in the isolated frog skin. J. Exp. Biol. 64, 587-602.

Sheehan M.E., Brauer, R.W. (1980) Implications of changes in temperature perception and temperature regulation in high pressure helium oxygen environments. In : Cox B., Lomax P., Milton A.S., Schönbaum E. (eds) Thermoregulatory mechanisms and their therapeutic implications. 232-237.

A EFFECTS OF HYDROSTATIC PRESSURE ON TRANSPORT AND OSMOREGULATION

EFFECTS OF HYDROSTATIC PRESSURE ON IONIC AND OSMOTIC REGULATION

R.D. ROER, A.J.R. PEQUEUX.

I. INTRODUCTION

The biological effects of hydrostatic pressure are extremely varied and depend not only upon the rate and magnitude of the pressure change, but also upon the organism and tissue being studied and the environmental (acclimatory) and evolutionary history of the animal.

Many pressure effects are not realized until pressures exceed that of an animal's habitat by some 300-3000 atm. Such effects include cell rounding in protozoa, marine eggs and cells in culture; disassembly of microtubules and, hence of ciliary and mitotic apparati; inhibition of DNA and RNA synthesis; and modification and denaturation of enzymes and other proteins (for a review see Zimmerman and Zimmerman, 1976).

Effects of much smaller changes in pressure (20-200 atm) have also been described. These are manifested in alteration of growth rates (Roer *et al.*, 1985), metabolic rate (Roer *et al.*, 1984b) and a variety of membrane associated phenomena.

Much of the research regarding the effects of changes in pressure on membrane-related processes has focused upon excitable tissues, attempting to elucidate the etiology of pressure-induced neurological disturbances (for a review see Wann and Macdonald, 1980). This research will be treated in other chapters in this volume. The present paper will address the effects of pressure on membrane-associated phenomena in non-excitable tissues, principally in regard to ionic and osmotic regulation in aquatic organisms.

II. PRESSURE EFFECTS ON TRANSPORT ACROSS CELLULAR AND EPITHELIAL SYSTEMS

A. Isolated tissues and cells

Brouha and co-workers (1970) demonstrated that the transepithelial potential difference (T.E.P.) across the isolated abdominal skin from the frog *Rana temporaria* was sensitive to an increase in hydrostatic pressure. A pressure of 97 atm (100 kg/cm^2) induced a hyperpolarization of the T.E.P. by as much as + 40mV. Ion substitution experiments pointed to an increase in Na permeability across the outer facing membrane as the source of the pressure-induced change in T.E.P. Further work on this system by Péqueux (1976a,b) utilizing short-circuit current measurements confirmed these findings and showed, moreover, that concomitant with the pressure-induced increase in passive Na influx, there was a stimulation of active Na transport. These effects were also shown to vary with the applied pressure : a reversible hyperpolarization of the T.E.P. was obtained with pressure : a reversible hyperpolarization of the T.E.P. was obtained with pressures of 48 to 484 atm (50-500 kg/cm^2), while pressures in excess of 581 atm (600 kg/cm^2) caused a prolonged depolarization which was irreversible following compression in the range of 774-968 atm (800-1000 kg/cm^2). Hyperpolarization was maximal in the vicinity of 194 atm (200 kg/cm^2).

The gills isolated from seawater-acclimated eels, *Anguilla anguilla*, also revealed an effect of pressure on passive Na permeability (Péqueux, 1972). The pressure effect here, however, was shown to be biphasic. An increase in hydrostatic pressure of 250 atm resulted in a 36% reduction of passive Na influx, while a pressure of 500 atm induced a 150% increase in Na influx.

The effect of pressure on Na transport in human red blood cells presents quite a different picture. Goldinger *et al.* (1980) demonstrated à 40% inhibition of active Na extrusion from erythrocytes by pressures of only 30 to 150 atm. These pressures resulted in increased intracellular levels of ATP, and thus it was concluded that the inhibition was a direct effect of pressure on the Na pump rather than an indirect effect of interference with glycolysis.

B. Whole animal transport

The dichotomy of pressure effects, on the one hand affecting permeability and on the other affecting active transport, is seen in studies of pressure effects on iono- and osmoregulation in intact animals as well as in the isolated tissues.

In the freshwater crayfish, *Procambarus clarkii*, the active uptake of Na is severely inhibited by increased hydrostatic pressures (Roer and Shelton, 1982). As little as 15 or 25 atm has an inhibitory effect on Na influx after one hour (Fig.1); 50 or 100 atm has an im-

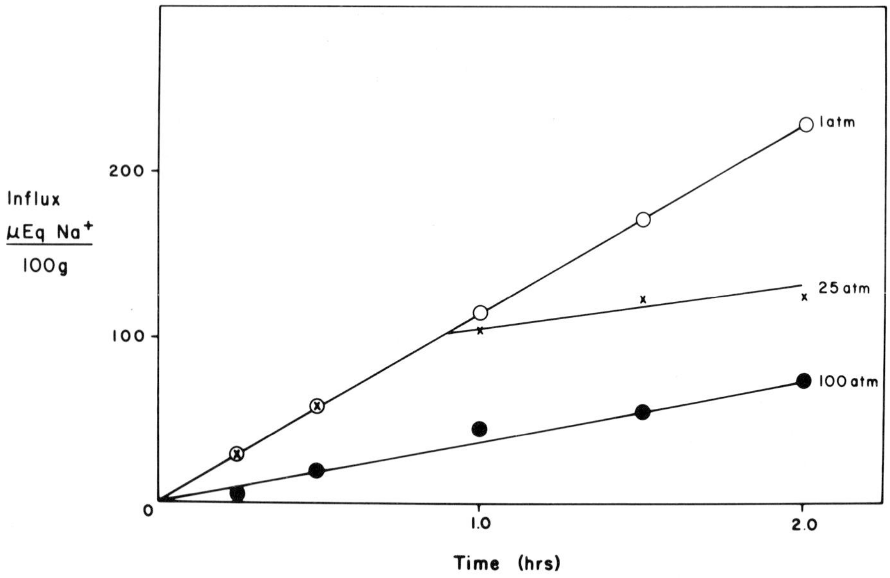

Fig.1. *Representative plots of unidirectional uptake of Na by crayfish (P. clarkii) at 3 different pressures. Influx rates are derived from linear regressions (After Roer and Shelton, 1982)*

mediate effect, both inhibiting influx by approximately 80%. These effects are reversible, but recovery requires 1-2 hrs at 1 atm. There is no effect of pressure on passive efflux.

Freshwater gammarids demonstrate a similar inhibition of active Na uptake in response to a change in hydrostatic pressure. In studies of populations of both deep- and shallow-dwelling species of gammarids from Lake Baikal in Central Siberia, U.S.S.R., it was found that both increases and decreases in pressure may have an effect on ionic regulation. As in the shallow-dwelling crayfish, the shallow-dwelling freshwater gammarids experience an approximate 35% inhibition of Na influx when exposed to 100 atm (Roer *et al.*, 1984a). Conversely a decrease in pressure to 1 atm from that at the habitat depth of deep-dwelling gammarids (60-140 atm) also results in some 40% inhibition of active uptake of Na (Roer *et al.*, 1984a; Brauer *et al.*, 1980). In neither case is there a demonstrable effect of pressure on passive Na permeability.

The effect of pressure on the euryhaline green crab, *Carcinus maenas*, is manifested by a different mechanism. Crabs were acclimated to 3 different salinities : 100%, 25% and 10% seawater for periods exceeding 2 weeks. Following acclimation, hemolymph {Na} averaged 254, 294 and 491 mM in the 10% (50 mM Na), 25% (125 mM Na) and 100% (480 mM Na) media respectively. Animals were then used for the measurement of unidirectional Na influx (according to the method of Roer and Shelton, 1982) and unidirectional efflux of Na. For efflux experiments, animals were injected with ∿10μCi of ^{22}NaCl in 100μl crustacean ringer solution (Roer, 1980) via the arthrodial membrane at the base of the 3rd pereiopod. Crabs were allowed to equilibrate for 2 hours after which time a hemolymph sample was drawn by syringe from the arthrodial membrane at the base of a pereiopod, and the animals were rinsed and placed in a 300 ml, well-stirred chamber for a 1.5hr. efflux at 1 atm; 100 μl samples of the external medium were removed in triplicate at 0, 15, 30, 45, 60 and 90 minutes for liquid scintillation counting. Following the 1 atm flux study, the crab was rinsed once again and placed in a well-stirred 300 ml chamber for an efflux measurement at 50 atm (Roer and Shelton, 1982). Compression was performed at a rate of 100 psi/min and sampling was performed as above. Following the 50-atm run, decompression was effected at 200 psi/min and the hemolymph of the animals was sampled once again. Flux rates were calculated based upon the rate of appearance of ^{22}Na in the medium and corrected for changes in specific activity over time and for the weight of the animal. All fluxes were performed at the salinity to which animals were acclimated.

Results of these studies demonstrated a decrease in the hemolymph Na concentration of animals acclimated to and tested at 25% and 10% seawater following compression to 50 atm (Fig.2). The {Na} of the 10% seawater-acclimated animals decreased by 14.3 ± 3.4%(N=5), that of 25% seawater-acclimated animals decreased by 10.0 ± 4.0%(N=5) (average ± S.E.M., $P < 0.05$ for both groups by Wilcoxon matched-pairs signed-ranks test). There was no significant change in the hemolymph {Na} of the 100% seawater-acclimated group (N=4).

The corresponding flux measurements revealed no demonstrable change in unidirectional Na influx at any salinity in response to pressure. There likewise was no significant change in the efflux of the 100% seawater-acclimated group on exposure to 50 atm. There was, however, an increase in the passive efflux of Na from crabs in dilute

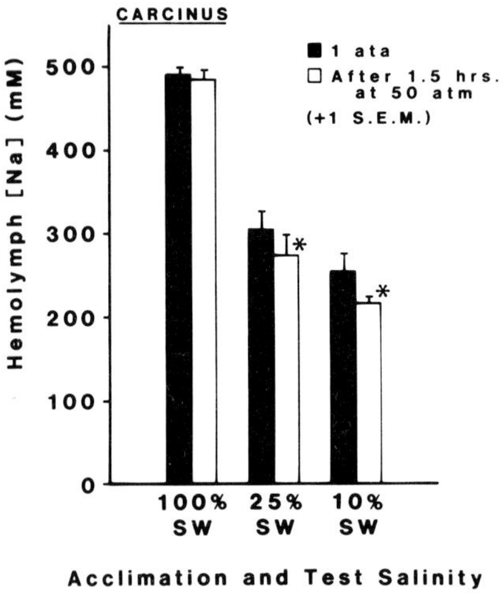

Fig.2. *Concentrations of Na in the hemolymph of green crabs (C. maenas) at 100, 25 and 10% seawater before (solid bars) and after (open bars) exposure to 50 atm for 1.5 hours. Vertical lines represent + 1 S.E.M. Asterisks denote a significant difference between pre- and postexposure levels at a given salinity (P <0.05 by Wilcoxon matched-pairs signed-ranks test).*

media upon compression to 50 atm (Fig.3). Pressure exposure increased efflux from 25% seawater acclimated crabs by 26.0 ± 16.7% (N=6) and from 10% acclimated crabs by 30.1 ± 23.2% (N=5) (Average ± S.E.M.; $P<0.05$ for 10% seawater animals by Levene's test and $P<0.05$ for both groups combined by Wilcoxon matched-pairs signed-ranks test).

C. Mode of pressure lability

Clearly, in all these systems thus far tested, pressures less than 300 atm affect the movement of ions across a membrane or series of membranes. In some systems the effect of pressure changes seems to be predominantly mediated by an alteration in permeability (top half of Table 1), while in others the effect is principally an inhibition of active transport (lower half of Table 1).

This dichotomy between the "permeability-labile" and the "transport-labile" systems may reflect the relative abilities of these properties to change with environmental conditions. The permeability-labile groups are all capable of altering the permeability of

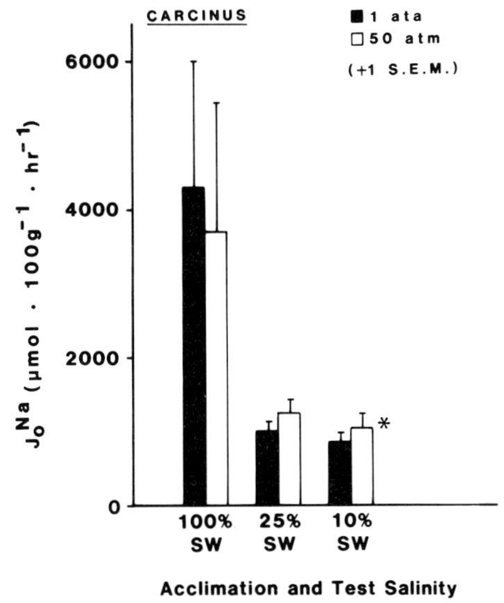

Fig.3. *Unidirectional efflux rates for sodium from green crabs (C.maenas) measured at 100, 25 and 10% seawater both at 1 atm (solid bars) and then at 50 atm (open bars). Vertical lines are + 1 S.E.M. Asterisk denotes a significant difference between rate measured at 1 atm and 50 atm at a given salinity (P<0.05 by Levene's test)*

their tissues in response to hormones and changes in salinity. This ability is most pronounced in *Carcinus* where the passive efflux of Na is very significantly reduced from 4300 µmol/100 g.hr in full seawater to 875 mol/100 g.hr in 10% seawater (P=0.008 by Mann-Whitney U-Test, P<0·025 by Levene's Test; Fig.3). This ability to alter permeability, and hence the susceptibility of this alteration to pressure, may not be present in the transport-labile group. The freshwater-adapted crustaceans have evolved a low salt and water permeability and may have lost the ability to alter it. It appears that the most labile component in the ion regulatory system is that which is affected by a change in pressure.

TABLE 1. Summary of pressure effects on permeability and transport in a variety of organisms and tissues

Tissue/organism	ΔPressure	Effect
Frog Skin[a] (*Rana temporaria*)	+48-290 atm	+10-40mV hyperpolarization of T.E.P. due to increase in Na permeability
	+581-968 atm	Inhibition of Na pump
Eel gill [b] (*Anguilla anguilla*)	+ 250 atm	36% reduction of passive Na influx
	+ 500 atm	150% increase in passive Na influx
Green crab (10% or 25% seawater) (*Carcinus maenas*)	+ 50 atm	25-30% increase in passive Na loss to medium
Red blood cells [d] (*Homo sapiens*)	+ 30-150 atm	40% inhibition of active Na efflux
Freshwater crayfish[e] (*Procambarus clarkii*)	+ 50-100 atm	80% inhibition of active Na uptake
Freshwater gammarids[f] (7 spp. from Lake Baikal)	+ 60-140 atm	30-40% inhibition of active Na uptake

Sources : a) Brouha *et al.*(1970); Péqueux (1976a)
b) Péqueux (1972)
c) Present work
d) Goldinger *et al.*(1980)
e) Roer and Shelton (1982)
f) Roer *et al.*(1984a)

While the immediate effect of pressure may be different in these two groups, the underlying pressure effect may be mediated by a common mechanism. Before discussing a working hypothesis for this common mechanism, it is worthwhile to review the effect of pressure on a major component of the Na transport system, the Na^+, K^+ - ATPase.

III. EFFECTS OF PRESSURE ON Na^+, K^+-ATPASE IN VITRO

A number of researchers have investigated the effect of hydrostatic pressure on the function of the isolated Na^+, K^+-ATPase *in vitro* with varying results. Péqueux (1976a) found no effect of pressure on frog skin Na^+, K^+-ATPase at 97 and 242 atm (100 and 250

kg/cm^2). At 484 atm (500 kg/cm^2), 40% inhibition was observed, and about 80% inhibition at 968 atm (1000/Kg/cm^2). Similar results were also obtained using Na^+, K^+ -ATPase preparations extracted from the gills and kidneys of the freshwater fish *(Ameiurus nebulosus)*, the seawater teleost *Cottus scorpius* and the seawater adapted eel *Anguilla anguilla* (Péqueux and Gilles, 1978). Moon (1975) found a progressive inhibition of Na^+, K^+-ATPase from salmon with increasing pressure, reaching 80% inhibition by 612 atm. The Na^+, K^+ -ATPase extracted from a specimen of abyssal fish, *Antimora rostrata*, captured at 1500 meters, on the other hand, showed activation up to 272 atm and then a progressive inhibition of activity with further increase in pressure (Moon, 1975).

These findings were confirmed in a more recent study by Pfeiler (1978) using Na^+, K^+-ATPase isolated from the gills of the shallow water fish *Scorpaena guttata* and the deeper living (300-1500 m) *Anoplopoma fimbria*.The enzyme from both fish showed activation by pressures up to 136 atm. Activation of the ATPase from *Anoplopoma* persisted up to a pressure of 250 atm. Thereafter, the Na^+, K^+-ATPases from both fish exhibited progressive inhibition with further increases in pressure up to 476 atm, at which point activity was still 60% of values at 1 atm.

These results on isolated ATPase systems are not consistent with our results using crayfish, wherein *in vivo* Na^+ transport is inhibited by nearly 90% with hydrostatic pressures of only 50 atm. A discrepancy of this nature was also evident in the work of Goldinger *et al.* (1980) in which active Na^+ extrusion from human red blood cells was inhibited by 40-50% at 50 atm, while Na^+, K^+ - ATPase isolated from these cells showed activation with pressures up to 150 atm.

Some of the disparity in results may arise from different methods for the isolation and treatment of the enzyme extract. As will be discussed below, the lipid milieu of the Na^+, K^+-ATPase is among a number of factors which strongly influence its catalytic activity.

IV. HYPOTHETICAL MECHANISM OF THE PRESSURE EFFECT

There are several possible sites of action for the pressure-induced alterations of permeability and active transport of Na. Pressure may act directly upon the membrane-bound proteins which facilitate these functions, affect the supply of counterions or acid-base balance through shifts in pH or affect the lipid milieu of the membrane in which the transport proteins function.

That pressure affects ion movements by alteration of lipid dynamics appears to be the most likely hypothesis, as effects of pressure on acid-base equilibria (Distèche, 1972; Péqueux, 1976b) and protein structure (Zimmerman and Zimmerman, 1976) generally are manifested at pressures far in excess of 100-200 atm. Both permeability (Bruner and Hall, 1983; Yorio *et al.*, 1983; Boheim *et al.*, 1980; Antonov *et al.*, 1980; De Kruijff and Van Zoelen, 1978) and the function of the Na^+, K^+ -ATPase (Kimelberg and Papahadjopoulos, 1972; Goldman and Albers, 1973; De Smedt *et al.*, 1979) depend heavily upon the physical state of the membrane phospholipid bilayer. It has been demonstrated, for example, that decreasing membrane fluidity via alterations in lipid composition, temperature, or pressure causes a decrease in Na^+, K^+ -ATPase activity (Sinensky *et al.*, 1979; Silvius and McElhaney, 1980).

A decrease in membrane fluidity and concomitant decrease in Na^+, K^+ -ATPase activity has been suggested as the mechanism by which increased hydrostatic pressure inhibits Na^+ transport in Crustacea (Roer and Shelton, 1982) and in prokaryotic cell membranes (MacNaughton and Macdonald, 1982). This hypothesis does not, however, account for the decrease in Na^+ influx in deep animals when the pressure is decreased from that at habitat depths. Such an effect of decreased pressure on Na transport is also indicated in the *in vitro* response of gill Na^+, K^+ -ATPase from the deep-sea fish, *Antimora rostrata* (Moon, 1975). Na^+, K^+-ATPase extracted from fish caught at 1500m showed activation with increasing pressure from 1 to 270 atm, followed by a dramatic decrease at higher pressures (See Fig.2 in Moon, 1975). A maximum or optimum point in the Na^+, K^+ - ATPase activity vs.temperature curve is also seen at different pressures in membrane-bound enzyme preparations from *Acholeplasma laidlawii* (See Fig.1 in Mac Naughtan and Macdonald, 1982), and in the conductance vs.temperature curves for gramicidin and alamethicin pores in synthetic bilayer membranes (Boheim *et al.*, 1980). In both cases, there is an increase in activity of the enzyme or conductance through the channel as the temperature is increased to the phase transition temperature, followed by a decrease in activity or conductance as the phase transition temperature is exceeded.

The fact that there is a maximum in the activity vs.pressure (or temperature) curves for these systems, with a decrease in activity as the pressure is either increased or decreased from some optimal value, indicates which aspects of the lipid physical state may be involved in the pressure effect on transport.If one considers not the static properties of the membrane lipids (such as structural order), but instead the dynamic properties which are proportional to the strength of the order parameter fluctuations, then functions emerge which can adequately describe the effects of both an increase and a decrease of pressure on transport. Static lipid order (s), measurable by deuterium magnetic resonance, electron spin resonance or fluorescence anisotropy, is a function of the uniaxial order of the acyl chains (Jähnig, 1981a). The static lipid order decreases monotonically as temperature decreases or as pressure increases with a sharp bend at the phase transition. The dynamic properties of the membrane, however, are a function of the strength of the order fluctuations (δS^2), which show a maximum at the phase transition. Such dynamic properties include coherence length, relaxation time, lateral compressibility and permeability (Jähnig, 1981a,b). Moreover, if membrane fluidity is considered not as a measure of structural disorder, but dynamically as a measure of structural lability, then fluidity, too, will be maximal at the phase transition (Jähnig, 1981b).

If we now think of the activity of membrane-bound enzymes as being dependent upon this "dynamic fluidity" , then the activity will be highest at the phase transition and will fall off as the physical state of the lipid becomes either more fully a liquid crystal or a gel. Thus, if the membrane lipids of an animal are poised at the phase transition under the conditions which obtain in its habitat, then an increase in temperature or a decrease in pressure should decrease the "dynamic fluidity" , as should a decrease in temperature or increase in pressure. Indeed, the membrane lipids of most animals studied are at the phase transition point at their acclimation temperatures and pressures (Jain, 1983).

V. PRESSURE ACCLIMATION

In that one generally views acclimation (or homeostasis) as the adaptive ability of an organism to adjust (or control) its internal milieu to maintain an acceptable steady state in the face of environmental change, one might assume that organisms would be incapable of acclimating to "exotic" environmental conditions - conditions never encountered in the evolutionary history of a species. We thus might well assume *a priori* that shallow-dwelling organisms which have not encountered high pressures (and whose ancestors had not encountered high pressures) would lack the ability to acclimate to large increases in pressure. This assumption, however, is based upon the misconception that an organism, faced with an environmental change, acclimates to that change *per se*. Rather, the organism acclimates, as we shall see, to the effects of the environmental change upon its internal milieu (Economos, 1982).

At the Institute for Marine Biomedical Research, we have developed the capacity to maintain animals for prolonged periods of time in a circulating high pressure aquarium system (Brauer and Royal, 1979; Roer *et al.*, 1985). Utilizing this transfer pump high-pressure aquarium system (TPHPA) we have conducted experiments on the ability of freshwater crayfish and green crabs to withstand long-term exposures to high pressures.

Measurements of hemolymph Na concentrations in *Carcinus* after varying periods of exposure to 50 atm in 25% seawater reveal that after the slight depression noted above following 1.5 hours at pressure, there is no significant depression of hemolymph Na levels after 3 hours to 21 days of exposure (Fig.4). Likewise, measurements of hemolymph Na concentrations in *Procambarus* both before and after exposure to 100 atm in water containing 0.4 mM Na reveal no changes from 3 to 24 hours in blood Na (Fig.5) or osmolality. Crayfish are capable of withstanding exposure to 100 atm in the TPHPA for at least 6 weeks.

The ability of these animals to live at these high pressures without any noticeable decrease in blood Na concentration must result from a rapid compensation and eventual acclimation to the effects of pressure on ion transport and permeability. The acclimation of trans-

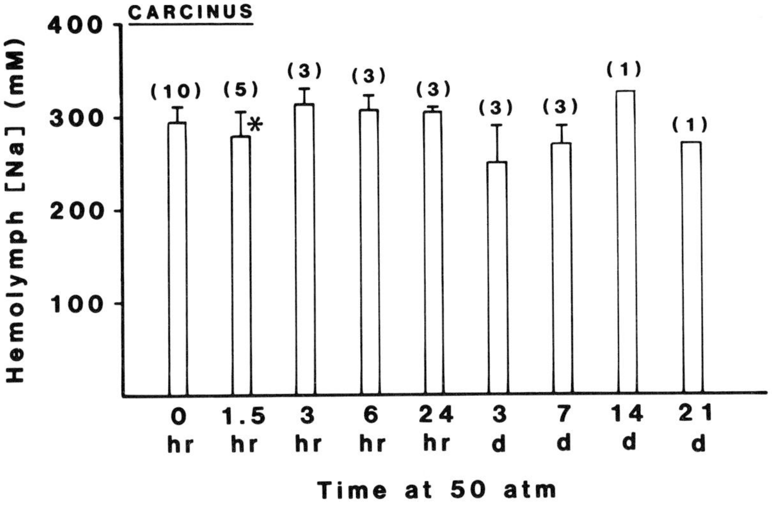

Fig.4. Hemolymph Na concentrations of green crabs (C.maenas) following various periods of exposure to 50 atm in 25% seawater. Number of animals tested are in parentheses. Vertical lines are + 1 S.E.M. Asterisk denotes significant difference from control (0 hours exposure) animals (P< 0.05 by Mann-Whitney U-test).

port function in crayfish is indicated in preliminary flux studies which demonstrate a positive shift in the net flux of Na in 100 atm-acclimated animals when tested against control animals at 100 atm (Fig.6).

This ability of crayfish to acclimate to pressure with respect to ion transport is in sharp contrast to the inability of juvenile crayfish to acclimate to 100 atm with respect to growth and molting (Roer *et al.*, 1985). This difference once again enforces the notion that acclimation occurs in response to specific effects of pressure, rather than to increased pressure *per se*.

The nature of the rapid compensation of these Crustacea to the ion transport alterations induced by pressure are, as yet, unclear. It is quite possible, however, and consistent with the hypothesis elaborated in Section 4, that the acclimatory response of these animals to long-term pressure exposure may entail changes in the membrane lipids.

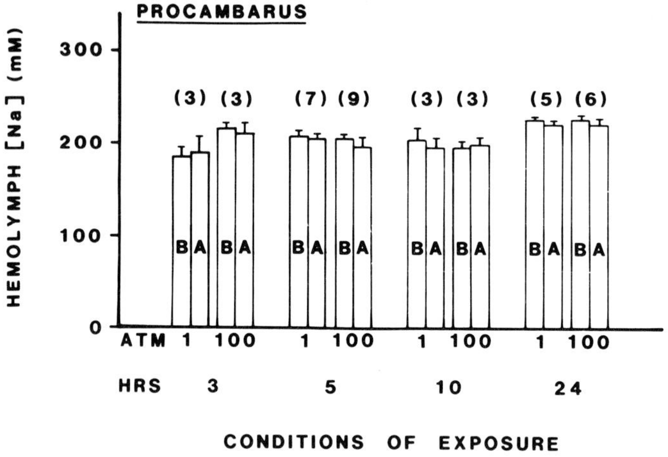

Fig. 5. *Hemolymph Na concentrations of crayfish (P. clarkii) before (B) and after (A) varying periods of exposure to 100 atm. Number of animals tested are in parentheses. Vertical lines are + 1 S.E.M.*

Many organisms are capable of altering the composition of their membrane lipids upon acclimation to different temperatures. These changes include the ratios of phospholipid classes (Chapelle *et al.*, 1979, 1982) , the degree of saturation in the acyl chains (Cossins and Prosser, 1978; Chapelle, 1978), and the degree of sialization of gangliosides (Hilbig *et al.*, 1979). It has been suggested (Hazel and Prosser, 1974; Cossins and Prosser, 1978) that such alteration in membrane lipids is not in response to the change in temperature, but rather in response to the temperature-induced alteration of membrane fluidity.

This notion of "homeoviscous" acclimation is consistent with the present data on pressure acclimation. Pressure , like temperature, is capable of affecting membrane fluidity and phase transitions. Acclimatory modifications of membrane composition may then occur in response to the altered dynamic fluidity of the membrane rather than as a direct consequence of pressure changes *per se*.

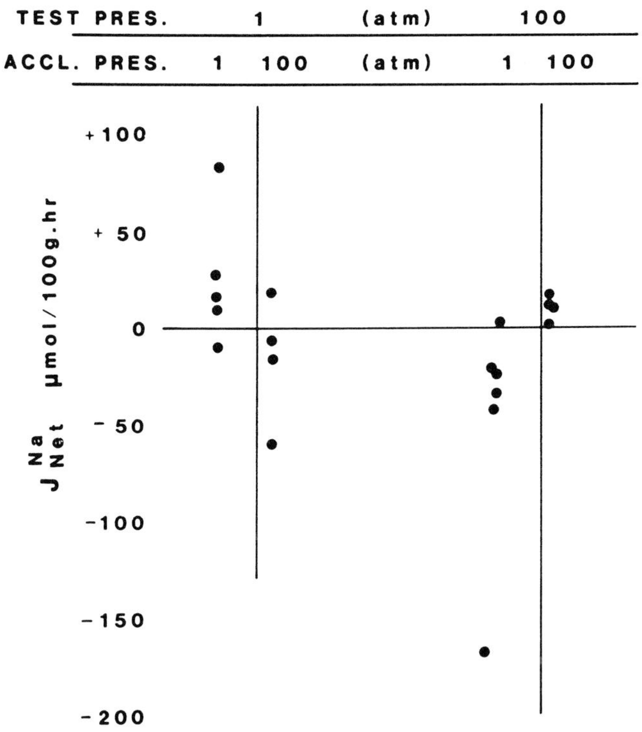

Fig. 6. Unidirectional influx rates for Na from crayfish acclimated for at least 18 days to 100 atm compared to 1 atm - acclimated controls, tested at 1 and 100 atm. Each point represents one measurement on an individual animal.

Acknowledgements. The work presented was supported, in part, by the NATO Science Fellowship Program, Belgian FNRS, US-Belgian Cooperative Science Program and the Griffis Foundation.

REFERENCES

Antonov, V.F., Petrov, V.V., Molnar, A.A., Predvoditelev, D.A., Ivanov, A.A. (1980) The appearance of single-ion channels in unmodified lipid bilayer membranes at the phase transition temperature. Nature, 283, 585-586.

Boheim, G., Hanke, W., Eibl, H. (1980) Lipid phase transition in planar bilayer membrane and its effect on carrier- and pore-mediated ion transport. Proc. Natl. Acad. Sci. USA, 77, 3403-3407.

Brauer, R.W., Bekman, M.Y., Keyser, J.B., Nesbitt, D.L., Shvetzov, S. G., Sidelev, G.N., Wright, S.L.(1980) Comparative studies of sodium transport and its relation to hydrostatic pressure in deep and shallow water gammarid crustaceans from Lake Baikal. Comp. Biochem. Physiol. 65A, 129-134.

Brauer, R.W., Royal, R.D.(1979) A system's approach to live retrieval of and experimentation with deep sea animals. I. The life support system. In : Ocean Energy Marine Technol. Society, Washington D.C., pp.65-69.

Brouha, A., Péqueux, A., Schoffeniels, E., Distèche, A.(1970) The effects of high hydrostatic pressure on the permeability characteristics of the isolated frog skin. Biochim. Biophys. Acta 219, 455-462.

Bruner, L.J., Hall, J.E.(1983). Pressure effects on alamethicin conductance in bilayer membranes. Biophys. J. 44, 39-48.

Chapelle, S. (1978) The influence of acclimation temperature on the fatty acid composition of an aquatic crustacean (*Carcinus maenas)*. J. Exp. Zool. 204, 337-346.

Chapelle, S., Brichon, G., Zwingelstein, G.(1982) Effect of environmental temperature on the incorporation of ^{3}H-ethanolamine into the phospholipids of the tissues of the crab *Carcinus maenas*. J. Exp. Zool. 224, 289-297.

Chapelle, S., Zwingelstein, G., Meister, R., Brichon, G.(1979). The influence of acclimation temperature on the phospholipid metabolism of an aquatic Crustacea (*Carcinus maenas*) J.Exp. Zool. 210, 371-380.

Cossins, A.R., Prosser, C.L.(1978) Evolutionary adaptation of membranes to temperature. Proc. Natl. Acad. Sci. USA 75, 2040-2043.

De Kruijff, B., Van Zoelen, E.J.J.(1978) Effect of the phase transition on the trans-bilayer movement of dimyristoyl phosphatidylcholine in unilamellar vesicles. Biochim. Biophys. Acta, 511, 105-115.

De Smedt, H., Borghgraef, R., Ceuterick, F., Heremans, K.(1979). Pressure effects on lipid-protein interactions in $(Na^{+}+K^{+})$-ATPase. Biochim. Biophys. Acta 556, 479-489.

Distèche, A.(1972). Effects of pressure on the dissociation of weak acids. Symp. Soc. Exp. Biol. 26, 27-60.

Economos, A.C. (1982) Human homeostasis in the space environment : A systems synthesis approach. In : Miquel, A., Economos, A.C. (Eds) Space Gerontology. NASA Conf. Pub. 2248 pp.22-23.

Goldinger, J.M., Kang, B.S., Choo, Y.E., Paganelli, C.V., Hong. S.K. (1980) Effect of hydrostatic pressure on ion transport and metabolism in human erythrocytes. J. Appl. Physiol. Respir. Environ. 49, 224-231.

Goldman, S.S., Albers, R.W. (1973) Sodium-potassium activated adenosine triphosphatase. IX. The role of phospholipids. J. Biol. Chem. 248, 867-874.

Hazel, J.R., Prosser, C.L. (1974) Molecular mechanisms of temperature compensation in poikilotherms. Physiol. Rev. 54, 620-677.

Hilbig, R., Rahmann, H., Rösner, H. (1979) Brain gangliosides and temperature adaptation in eury- and stenothermic teleost fish (carp and rainbow trout). J. Therm. Biol. 4, 29-34.

Jähnig, F. (1981a) Critical effects from lipid-protein interaction in membranes. I. Theoretical description. Biophys. J. 36, 329-345.

Jähnig, F. (1981b) Critical effects from lipid-protein interaction in membranes. II. Interpretation of experimental results. Biophys. J. 36, 347-357.

Jain, M.K. (1983) Nonrandom lateral organization in bilayers and biomembranes. In : Aloia, R.C. (ed) Membrane fluidity in biology vol I. Academic Press, New York. pp.1-37.

Kimelberg, H.K., Papahadjopoulos, D. (1972) Phospholipid requirements for $(Na^{+}+K^{+})$-ATPase activity : Head group specificity and fatty acid fluidity. Biochim. Biophys. Acta 282, 277-292.

MacNaughtan, W., Macdonald, A.G. (1982) Effects of pressure and pressure antagonists on the growth and membrane-bound ATPase of *Acholeplasma laidlawii* B. Comp. Biochem. Physiol. 72A, 405-414.

Moon, T.W. (1975) Effects of hydrostatic pressure on gill Na-K-ATPase in an abyssal and a surface-dwelling teleost. Comp. Biochem. Physiol. 52B, 59-65.

Péqueux, A. (1972) Hydrostatic pressure and membrane permeability. The effects of pressure on organisms. Symp.Soc.Exp.Biol. 26, 483-484.

Péqueux, A.(1976a) Polarization variations induced by high hydrostatic pressures in the isolated frog skin as related to the effects on passive ionic permeability and active Na^+ transport. J. Exp. Biol. 64, 587-602.

Péqueux, A.(1976b) Effects of pH changes on the frog skin electrical potential difference and on the potential variations induced by high hydrostatic pressures. Comp. Biochem. Physiol. 55A, 103-108.

Péqueux, A., Gilles, R.(1978) Effects of high hydrostatic pressures on the activity of the membrane ATPases of some organs implicated in hydromineral regulation. Comp. Biochem. Physiol. 59B, 207-212.

Pfeiler, E.(1978) Effects of hydrostatic pressure on $(Na^+ + K^+)$-ATPase and Mg^{2+}-ATPase in gills of marine teleost fish. J. Exp. Zool. 205, 393-402.

Roer, R.D.(1980) Mechanisms of resorption and deposition of calcium in the carapace of the crab *Carcinus maenas*. J. Exp. Biol. 88, 205-218.

Roer, R.D., Bekman, M.Y., Shelton, M.G., Brauer, R.W., Shvetzov, S.G. (1984a) The effects of changes in hydrostatic pressure on Na transport in gammarid amphipods from Lake Baikal. J. Exp. Zool. 232(2). In press.

Roer, R.D., Dillaman, R.D., Shelton, M.G., Brauer, R.W.(1985) Effects of pressure on growth. In : Wenner, E.(ed) Crustacean Issues 2 : Crustacean growth. AA. Balkema Publ., Rotterdam. In press.

Roer, R.D., Shelton, M.G.(1982) Effects of hydrostatic pressure on Na transport in the freshwater crayfish, *Procambarus clarkii*. Comp. Biochem. Physiol. 71A, 271-276.

Roer, R.D., Sidelyova, V.G., Brauer, R.W., Galazii, G.I.(1984b) Effects of pressure on oxygen consumption in cottid fish from Lake Baikal. Experientia 40, 771-773.

Silvius, J.R., McElhaney, R.N.(1980) Membrane lipid physical state and modulation of the Na^+, Mg^{2+}-ATPase activity in *Acholeplasma laidlawii* B. Proc. Natl. Acad. Sci. USA 77, 1255-1259.

Sinensky, M., Pinkerton , F., Sutherland, E., Simon F.R.(1979) Rate limitation of $(Na^+ + K^+)$-stimulated adenosinetriphosphatase by membrane acyl chain ordering. Proc. Natl. Acad. Sci. USA 76, 4893-4897.

Wann, K.T., Macdonald, A.G.(1980) The effects of pressure on excitable cells. Comp. Biochem. Physiol. 66A, 1-12.

Yorio, T., Torres, S., Tarapoom, N.(1983) Alteration in membrane permeability by diacylglycerol and phosphatidylcholine containing arachidonic acid. Lipids 18, 96-99.

Zimmermann, A.M., Zimmerman, S.(1976) Influences of hydrostatic pressure on biological system. In : Lambertsen C.J.(ed) Underwater Physiology vol. V. FASEB, Bethesda Md. pp.381-396.

IONIC REGULATION UNDER HIGH PRESSURE : SOME OBSERVATIONS OF COMPARATIVE AND THEORETICAL INTEREST

A.G. MACDONALD, C.J. SHELTON.

I. INTRODUCTION

Understanding the adaptations of cells to deep sea conditions clearly requires knowledge of how transport processes are effected by high pressure. The use of high pressure as an analytical variable in chemistry and biochemistry has, from time to time, prompted its extension to very complex processes, such as the active transport of solutes across cell membranes. In both areas of research the pressure range of interest is about 1000 atm or 100 MPa, corresponding to the pressure occurring at 10,000m depth. This is close to the greatest depth in the oceans (10790m in the Marianas Trench). A more useful index of the pressure and depth distribution of the oceans is 3800m, their mean depth. Approximately 50% of our planet lies beneath this, or at a greater depth of water.

With the exception of high pressure the deep sea environment comprises a relatively familiar set of conditions. The ionic composition of deep sea water is essentially the same as that of the shallow sea; the dissolved gases are similar to those in shallow water, present at the usual concentrations but at slightly increased activities or fugacities, due to the high hydrostatic pressure (see below). The temperature is low (<4°C) and constant, and food for microorganisms and animals alike is in short supply (Ernst & Morin, 1982; Macdonald, 1975).

The general expectation then, is that adaptations to high pressure might have to economise on energy and make the most of the ambient low temperature and high pressure, each of which contributes to the free energy change in chemical reactions.

$$\Delta G = \Delta E + P.\Delta V - T.\Delta S \qquad \text{(i)}$$

It is through the $P.\Delta V$ term that high pressure affects equilibrium and rate constants and causes profound physiological and biochemical disruption when applied to shallow water organisms.

Exceptional conditions prevail in the spectacular hydrothermal vents in the Pacific Ocean, which are oases in the cold desert of ocean floor through which hot brines pour, and around which a flourishing flora and fauna develops (Rona *et al.*,1984).

The purpose of this paper is to discuss some selected problems associated with studying transport processes under high pressure. This is done under two headings, practical aspects and theoretical aspects. No attempt is made to systematically review any of the constituent subjects which, in any case, have been dealt with in a number of recent specialised reviews (Morild, 1981; Heremans, 1982; Weber and Drickamer, 1983; Macdonald, 1984).

II. PRACTICAL ASPECTS

Experiments involving a cell suspension or tissue preparation in a pressure vessel involve awkward constraints and sometimes difficult technical problems. In this section we describe some of the problems and their solution which arose in a study of ion transport in fish erythrocytes under pressure.

A. Fish erythrocytes

It is of considerable interest to establish how high hydrostatic pressure affects the better understood transport processes in ordinary fish erythrocytes, because of the prospects for eventually studying the pressure tolerance of erythrocytes from deep sea fish (Shelton *et al.*,1984). In general, nucleated erythrocytes possess co- and counter-transport systems which play a part in non steady-state volume regulation (Kregenow, 1981; Cala, 1983). In fish erythrocytes this system accounts for about one half of the net ionic flux even when the cell volume is unperturbed. Cellular ionic regulation is further complicated in some nucleated erythrocytes by the occurrence of catecholamines-stimulated swelling (Riddick *et al.*, 1971; Nikinmaa, 1982). This occurs with a concomitant increase in the unidirectional ion fluxes. In fish erythrocytes hormone sensitive ion transport has been a complicating factor in studies of the effect of other variables on membrane ion transport, for example temperature (Bourne and Cossins, 1982). Stress imposed on the fish during the blood sampling raises the concentration of plasma catecholamines sufficiently to increase the ion fluxes across the membrane and to cause the cells to swell. These effects diminish within a period of hours if the plasma is washed off the cells and replaced with an artificial medium (Bourne and Cossins, 1982; Shelton and Macdonald, unpublished). Also, the effects of plasma catecholamines may be blocked with the β adrenergic antagonist propanolol (Riddick *et al.*, 1971; Shelton and Macdonald, unpublished data). Clearly fish erythrocytes are complex and delicate cells to work with, but as "free living" unicells they are relatively convenient for measurements of transmembrane fluxes under high pressure.

B. Methods for measuring ion fluxes at high hydrostatic pressure

We used conventional methods, modified to enable the cells to be incubated at high pressure. Suspensions of erythrocytes were incubated in 5 ml plastic syringes which were pressurised in a 70mm bore stainless steel vessel capable of withstanding 1000 atm internal pressure. The vessel was pressurised by a hand operated pump connected via a flexible capillary tube which allowed the vessel to be rocked through an arc of 180° to stir the cells within the syringes. The vessel was also equipped with an end plug which could be removed rapidly and its temperature was controlled by externally circulated water. Water transmitted pressure to the syringes, and in so far as the suspension of erythrocytes contained no gas bubbles, the cells were subjected to increased hydrostatic pressure alone, with the activity (fugacity or partial pressure) of dissolved gases only changing as a function of hydrostatic pressure i.e.

$$fO_2 \triangleq PO_2 . \exp \frac{VO_2 . P}{RT}$$

(Taylor, 1978). In aqueous solvents these changes are exponential at low pressure, approximately $\pm$ 14%/100 atm^{-1} for oxygen. The selected hydrostatic pressure was achieved within 10 sec after sealing the vessel. Rapid decompression enabled an aliquot of erythrocytes to be withdrawn from the syringe, and pressure was restored within 90 seconds of decompression. Changes in the temperature of the water arising from changes in pressure are negligible against the constant temperature of 10 $\pm$ 0.2°C used in the experiments.

Erythrocytes from the plaice *Pleuronectes platessa* were used, with the objective of measuring the fluxes of ^{86}Rb and ^{22}Na under pressure. Full details of the experimental procedures will be given elsewhere (Shelton and Macdonald, unpublished data). Briefly, it was necessary to stabilise the cells by incubating them in experimental medium for five hours before use, after which no further changes in ion fluxes were observed. Influx and efflux curves were obtained in the usual way. For example, ^{22}Na efflux was measured in cells which had been exposed to the isotope in experimental medium for four hours. After washing in fresh experimental medium the ^{22}Na efflux was measured after 40 minutes incubation at pressures up to 400 atm. Fig.1 shows the efflux to be linear under pressure.

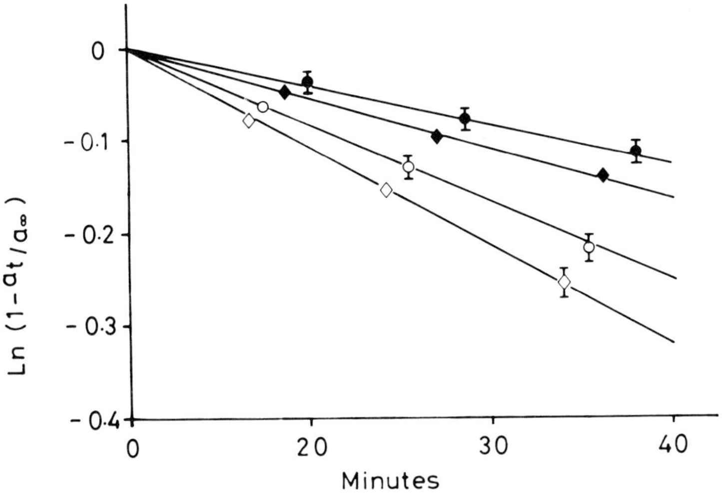

Fig.1. ^{22}*Na efflux from plaice erythrocytes at 300 atm, 10°C in the presence and absence of* 10^{-4}*M ouabain. Each point represents the mean of 3 determinations on one sample of erythrocytes ± S.D.(The line is drawn using the least squares method). The cells had previously been incubated in experimental medium for 5 hours to stabilize volume regulating ion fluxes. The medium contained NaCl, 200mM; KCl, 5mM;* $CaCl_2$*, 1mM;* $MgCl_2$*, 1mM; Glucose, 5 mM; Bovine serum albumin, 10g* l^{-1}*; Tris buffer, 10 mM, pH 7.75 at 10°C.* a_t *is the external* ^{22}*Na activity at time t and* a_∞ *the external activity at isotopic equilibrium. The efflux constant* $kt = ln(1 - \frac{at}{a_\infty})$ *(Brahms, 1977).*

◇*, 1atm;* ◆*, 1atm plus ouabain;* o*, high pressure;* •*, high pressure plus ouabain.*

To obtain the Na efflux in absolute units (mM.PCV_*^{-1}. h^{-1})it is necessary to multiply the efflux rate constant (i.e. the slope of the line in Fig.1) by the concentration of Na inside the erythrocytes, {Na}i. Ideally {Na}i should be measured at pressure but in practice determinations have only been carried out on cells before or after incubation at pressure (as in the experiments with human erythrocytes under pressure : Hall *et al.* 1982; Goldinger *et al.*1980; Péqueux *et al.* 1980). Such a procedure is liable to error most likely from the

---

* P.C.V. = Packed Cell Volume.

movement of water across the cell membrane, causing significant changes in {Na}i. Haematocrit determinations carried out both before compression and after decompression provide an indication of any volume changes taking place, but they do not actually measure the volume of the cells whilst at high pressure. This was achieved in the present study by a modified Coulter counting apparatus (Fig.2). Coulter counting at high pressure was first carried out to measure cell multiplication (Macdonald 1967) and the apparatus illustrated in Fig.2 is a revised version of the earlier equipment. Coulter counting is based on the transient resistance change which occurs when a particle in an electrolyte solution (an erythrocyte in this case) passes through a small (70μ) hole in a glass tube. Electrodes on either side of the hole detect the changes in the resistance of the solution filling the hole. Within limits, certainly met in the present case, the resistance changes are proportional to the volume of the particles, and well established circuitry measures both the number and the sizes of the erythrocytes in suspension. The high pressure version works by a reciprocating piston forcing the cell suspension to and for across the orifice tube. The shaft driving the bevel gear, crank mechanism and piston actuates switches situated outside the pressure vessel which start and stop the electronic counting process. Counting starts as the crank passes through an arc of 40° , half way through the upon downstroke, when the velocity of the cell suspension passing through the orifice is both relatively constant and appropriate for the electronic counting process.

The reproducibility of the counting and volume determinations at different pressures was checked with latex particles (Fig.3). Volume determinations were carried out on plaice erythrocytes at normal atmospheric pressure, 10°C, and after 40 minutes incubation at 200 and 400 atm (Fig.3). No significant volume changes could be detected, so it is exceedingly likely that the values of {Na}i obtained at normal atmospheric pressure are very close to the actual values prevailing at high pressure. It seems most unlikely that ion fluxes occur sufficiently rapidly to cause serious error on the time scale of the experimental manipulation. Accordingly we have considerable confidence in these data.

The fluxes thus determined (Na efflux and ^{86}Rb influx) comprise at least three components : ouabain-inhibited, furosemide-inhibited and the balance which is probably mostly a simple diffusional leak. Their analysis will be discussed elsewhere (Shelton and Macdonald,

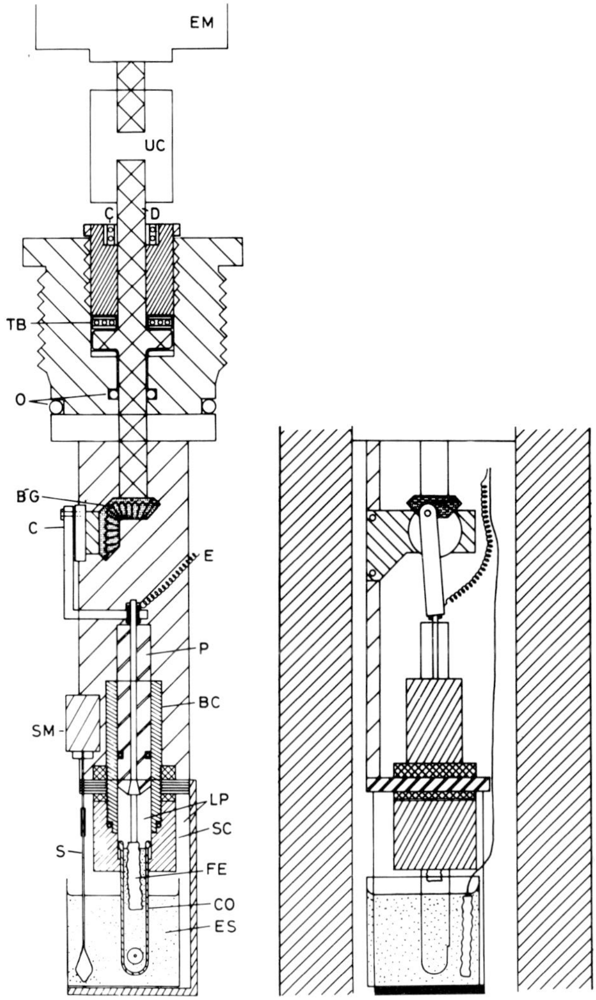

Fig.2. Coulter counting apparatus for measuring the volume of erythrocytes at pressures up to 700 atm. EM, electric motor which turns the drive shaft D through a universal coupling, UC, at approximately 1 RPM. C, centre bearing, TB, thrust bearing, O, O-ring seals, which seal the drive shaft and the end plug of the pressure vessel. The apparatus below the level of the end plug O-ring is shown in two sections (a) and (b) drawn at 90° to each other. In (b) a cross section of the pressure vessel is shown, bore = 5.8 cm. Temperature is controlled by means of water, circulating through an external copper coil (not shown). BG, bevel gear; C, crank; E, electrical lead connecting the foil electrode FE, which is mounted within the Coulter counter orifice tube, CO, to a high pressure connector in the end plug , not shown. An external electrode is shown in (b). The entire pressure vessel and counting apparatus is filled with liquid paraffin, LP, except for the beaker and orifice tube, CO, which contain the erythrocyte suspension (30ml). The orifice tube is sealed to a screwed cap SC, which fits to a brass cylinder, BC. A perspex piston, P, reciprocates within the cylinder, driven by the crank, C. An electric stirring motor, SM, drives a stirring paddle intermittently. The apparatus is shown with the crank close to top dead centre.

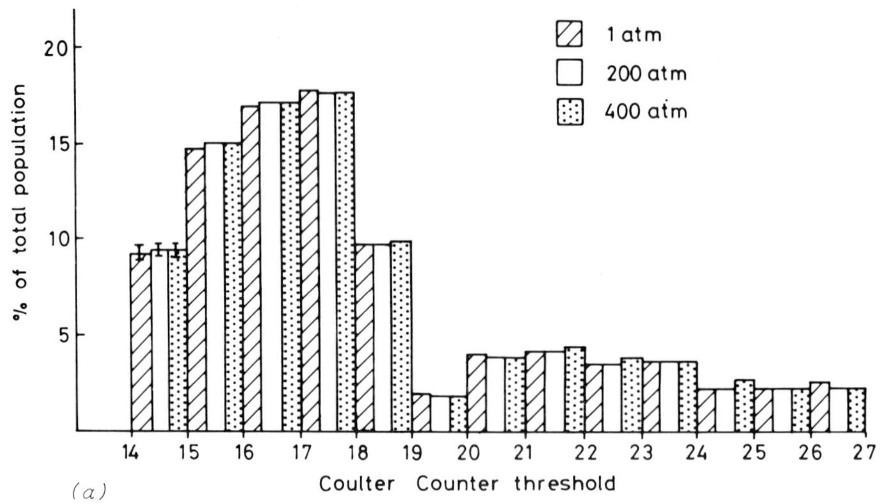

(a)

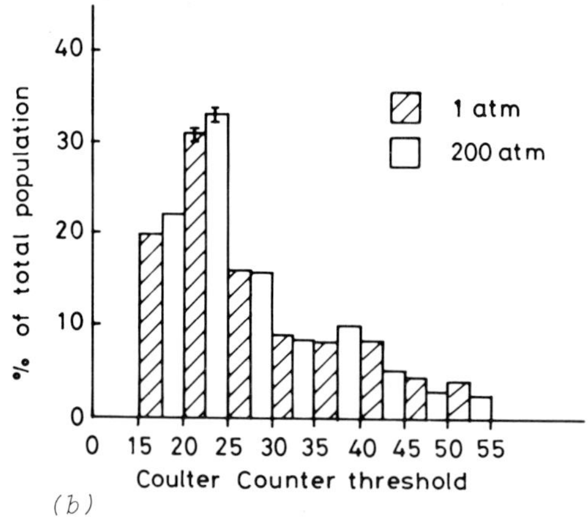

(b)

Fig.3. *Measurements of the volume of plaice erythrocytes at high pressure using a modified Coulter Counting apparatus. (a) control counts of latex particles (Median diameter = 4.86 µm) at different pressures. Model F Counter setting; aperture 1.6 , attenuation 1.0. (b) and (c) the size distribution of plaice erythrocytes in experimental medium at 200 and 400 atm, 10°C, respectively. The cells were first pressurised to 200 atm in increments of 50 atm at 15 min. intervals, incubated for 40 min., after which they were counted and their size distribution was determined. This took a further 20 min. The cells were then raised to 400 atm for a further 40 min and measured. Each column represents the mean of 4 counts ± S.D. Individual counts were typically 6X10^4 ml^{-1}. The counts at normal atmospheric pressure were carried out before pressure was applied.*
Haematocrit determinations before compression and post-decompression showed no difference.

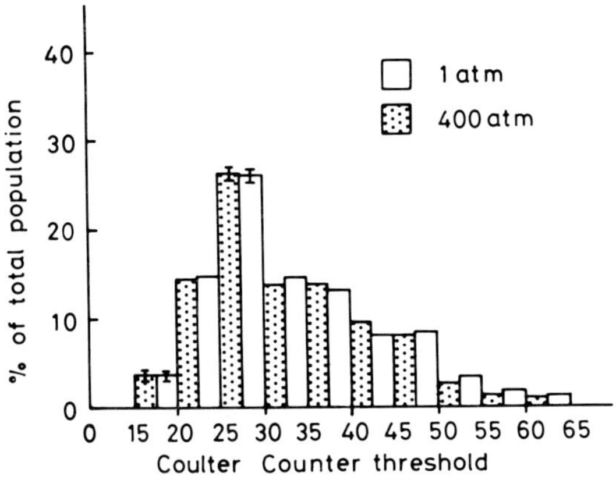

Fig. 3c

unpublished data) and here we simply note that pressure affects the different components in different ways. The ouabain-sensitive influx and efflux were inhibited by 10% and 25% respectively at 400 atm. The ouabain and furosemide insensitive influx of ^{86}Rb was increased by 100% at 400 atm. In both cases the plaice erythrocyte is similar to the human erythrocyte (Hall *et al.* 1982). The problems of interpreting this type of kinetic data are formidable and a selection of points of general interest are considered in the subsequent section.

III. THEORETICAL ASPECTS

A. Introduction

The analysis of the kinetic effects of high pressure is generally based on rate process theory. This states that the free energy of activation, $\Delta G^{\neq}$, of the rate-determining step is the essential phenomenon, comprising three constituents :

$$\Delta G^{\neq} = \Delta E^{\neq} + P.\Delta V^{\neq} - T.\Delta S^{\neq} \qquad \text{(ii)}$$

In complex reactions of the type considered here it is unlikely that a single rate-limiting step exists, but it is worth noting that in simple enzyme reactions, rate-limiting steps may be part of the main reaction or a side reaction. For example, the conversion of A to B in equation (iii) is the main reaction, but the catalysis may be rate determined by the side equilibrium between the enzyme and its ligand. A positive $\Delta V^{\neq}$ in the rate-determining step increases $\Delta G^{\neq}$ and hence reduces the rate of reaction under high pressure; a negative $\Delta V^{\neq}$ will cause the reaction to accelerate under high pressure

$$\text{A} \xrightarrow{\hspace{2cm}} \text{B} \qquad \text{(iii)}$$

$$\text{enzyme} \rightleftharpoons \text{ligand}$$

In physiological conditions (i.e. in the 1000 atm pressure range), $\Delta V^{\neq}$ is most likely to arise from solute-solvent interactions, and in the hypothetical example in equation (iii) $\Delta V^{\neq}$ could arise from hydration changes in the ligand-enzyme binding equilibrium, or perhaps from partial molar volume changes in the formation of the transition state complex as A becomes converted to B. With these points in mind the following theoretical section discuss solute-solvent interactions which are likely to be important to understanding the kinetics of transport processes under pressure.

B. Ion-water interactions

One widely held view of ions in aqueous solution is that their hydration layer comprises an inner dense electrostricted region and an outer layer whose local density is less than that of bulk water. High pressure favours a reduction in the outer, voluminous hydration layer, reducing the effective diameter of the ion in aqueous solution.

This is the likely explanation for the way pressure increases such parameters as specific ion conductance. For example at 5°C the relative specific conductance of KCl at infinite dilution is increased 10% by 1000 atm. Generally however such pressures exert only small effects on ionic properties in aqueous solutions. According to Nakahara (1972) the hydration numbers of K^+ and Cl^- do not change over the 1-1000 atm range, although at higher pressures K^+ shows a slight increase and Cl^- a slight decrease. Neilson (1979) argues that the average number of water molecules associated with Cl^- ions (in 3M solution) decreases from 5.5 to 4.4 over the 1-1000 atm range. Despite the difficulty of comparing these various indexes of hydration (which are essentially defined by the particular method of measurement) it is clear that ion-water interactions alone are unlikely to be sufficient to account for the effects which pressures of only several hundred atmospheres exert on various transport processes. Interactions with components of the membrane have to be considered, which is a familiar concept (see the review on selectivity sequences by Eisenmann and Horn 1983).

Consider now the example of the passive flux of K^+ through the human erythrocyte membrane. Under pressure the flux is accelerated (Hall *et al.* 1982) and it is interesting that the equivalent ouabain + furosemide insensitive flux in fish erythrocytes is also accelerated by pressure (section II,B). The apparent $-\Delta V^{\neq}$ in the rate-limiting step in the diffusion of K^+ through the human erythrocyte membrane appears to be reduced by substituting Cl^- ions with Br^-, NO_3^- or I^- ions. Thus the pressure-enhanced flux of K^+ is almost abolished by replacing Cl^- with I^- ions (Hall *et al.* 1982). As discussed elsewhere (Macdonald 1984) the most simple yet plausible reaction scheme is :

$$\text{hydrated } K^+ + \text{hydrated channel} \rightleftharpoons \text{K-channel}^{\neq} \rightarrow \text{K-channel} + \text{water}$$

$$\text{K-channel} \rightarrow \text{hydrated } K^+ + \text{hydrated channel} \qquad \text{(iv)}$$

The anions in question could affect $\Delta V^{\neq}$ by altering the structure of the bulk solution, but anion substitutions in certain enzyme reactions under high pressure have been shown to exert complicated effects which cannot be ignored here. According to Low and Somero (1975) in reactions involving lactate dehydrogenase or pyruvate kinase at high pressure, anion substitutions affect the rate of reaction by a mechanism additional to their effect on $\Delta V^{\neq}$. From equation (ii) it is clear that $\Delta E^{\neq}$ and/or $\Delta S^{\neq}$ must also be involved. The molecular picture proposed by the authors is that the enzyme's catalytic function involves some unfolding and exposure of amino acid side chains to the solvent water. Volume changes arise from these and anion substitutions may well modify the resultant $\Delta V^{\neq}$, but changes in $\Delta E^{\neq}$ and $\Delta S^{\neq}$, which are tightly coupled to the conformational changes underlying the $\Delta V^{\neq}$, also occur, contributing to a change in $\Delta G^{\neq}$ greater than that caused by $P.\Delta V^{\neq}$ alone. In other words, $\Delta E^{\neq}$ and $\Delta S^{\neq}$, assumed to be constant in the preamble to this section, may vary. It is therefore important to stress that the interpretation of the way pressure accelerates the passive K^+ flux in human erythrocytes assumes $\Delta E^{\neq}$ and $\Delta S^{\neq}$ to be constant. The only firm conclusion is that much more detailed knowledge of transport pathways in general is necessary for a rigorous analysis of the effects of pressure to be undertaken. This may well involve reconstituting membrane components in model membranes and studying their properties in simplified conditions. Comparison with pressure-adapted membrane transport components from deep sea fish cells might eventually be possible.

C. Interactions with membrane lipids

Membrane bound enzymes, ionophores and channel proteins are effectively dissolved in the lipid bilayer common to most membranes. The bilayer's molecular motion is reduced by high pressure, a phenomenon which has been reviewed recently (Macdonald, 1984; see also Lakowicz & Thompson, 1983; Chong & Cossins, 1983, 1984 and Taniguchi *et al.* 1982). Attempts have been made to account for the effects of pressure on membrane transport components in terms of bilayer fluidity. One example is the ATP-ase activity of dog kidney medulla which is reduced by high pressure in a way which is closely matched by the reduction in the fluidity of the bilayer supporting the enzyme. In this case bilayer fluidity was measured by the fluorescence depolarization of DPH (Chong *et al.* 1982). Interestingly the same membrane-enzyme preparation may undergo a depletion of its cholesterol content, presumably increasing its fluidity, with little effect on activity (De Pont *et al.* 1983). However the activity of the ATP-ase in isolated membrane fragments is not to be equated with ion transport. In human erythrocytes the activity of the ATPase and the transport of Na^+ are differentially affected by pressure (Goldinger *et al.* 1980).

Attempts to relate bilayer fluidity with ionophore-mediated

transport under high pressure have met with some success. The valinomycin-mediated flux of K^+ through unilamellar liposome bilayers of specific compositions studied by Johnson & Miller (1975) is one example. Valinomycin is a circular molecule which, when bound to K^+ adopts a rigid structure. Binding involves the dehydration of the ion and the formation of hydrogen bonds with carbonyl oxygens which replace water molecules. It is the 6 oxygens which impose the selectivity for K^+. The exterior of the ionophore comprises hydrophobic groups which enable it to partition into the bilayer lipid (Dobler 1981). $\Delta V^{\neq}$ for K^+ through phosphatidylcholine - 6% phosphatidylserine bilayers was 38 ml $mole^{-1}$ ion transported at 23°C, and a similar value was obtained for bilayers incorporating 2% phosphatidic acid. In the absence of valinomycin the $\Delta V^{\neq}$ for K^+ flux was approximately half, but still positive in sign, (the opposite to the passive K^+ flux through the erythrocyte membrane - see section III,B). The question is, what determines $\Delta V^{\neq}$? The values are too big to be accounted for simply as a dehydration of the ion, so presumably ionophore-lipid interactions are involved. These may be of two types one involving headgroup hydration and the other kink-isomer volume fluctuations in the acyl chains (Trauble, 1971).

The ionophore lasolocid A is a polyether antibiotic which transports cations across bilayers by forming a dimer with a hydrophilic interior and a hydrophobic exterior. Its transport of Gd^{3+} ions across egg lecithin bilayers has been measured under pressure by a technique based on electron paramagnetic resonance (Balschi *et al.*, 1982). $\Delta V^{\neq}$ was 140 ml mol^{-1} ion transported, over the 1-1000 atm range. As with the previous example the authors recognized two sources of volume change, one involving dehydration of the ion and the other the ionophore-lipid interaction , but their analysis proceeded no further.

The behaviour of voltage-sensitive ion channels in membranes under high pressure has been studied somewhat indirectly in neurons (reviewed in Macdonald 1984, Wann & Macdonald, 1980) and more directly in the case of alamethicin in artificial bilayers (Bruner & Hall, 1983). As with the ionophore experiments the interpretation of the $\Delta V^{\neq}$ thus obtained is a sticking point at present. This raises the question, have the effects of high pressure on <u>any</u> sort of membrane component yet been interpreted as an interaction with bilayer lipids? The answer is no, but perhaps the example which is closest to that goal is the effect of pressure on bacteriorhodopsin in the membrane of *Halobacterium halobium* A constituent of the molecule, retinal, undergoes a rapid light-driven isomerization and subsequent relaxation, the kinetics of which exhibit a discontinuity in an Arrhenius plot, indicating a phase transition at -42°C, normal pressure (Tsuda *et al.*, 1983). When the bacteriorhodopsin is incorporated in dimyristyl phosphatidylcholine liposomes, the discontinuity occurs close to the phase transition temperature. The relaxation reaction in the model membranes proceeds with a $\Delta V^{\neq}$ of 42 ml $mole^{-1}$ below the transition temperature, and close to nil above the transition temperature. This implies the reaction is constrained by the highly ordered lipids in the former but not in the latter condition.

The membrane of *Halobacterium* is, of course, unusual, consisting of a protein lattice, with only half the normal amount of lipid present. It is therefore of particular interest that the effect of pressure on the Arrhenius discontinuity temperature (T) is very different to that in the model membrane: in the native membrane T increased by 0.053°C atm^{-1} and in model membrane by 0.024°C atm^{-1}. The latter is the same as the effect of pressure on the transition temperature of pure lipid bilayers, consistent with $\frac{dT}{dP} = \frac{\Delta V.T}{\Delta H}$ (Heremans, 1982; Macdonald, 1984), whereas 0.053°C is about twice the value for any known bilayer. ΔV arises from the "melting" of the acyl chains in the lipid molecules, and it appears that in the *Halobacterium* membrane the lipid is organized in a peculiar way, giving a high ratio of $\frac{\Delta V}{\Delta H}$. This example also illustrates two practical points which are relevant to the study of transport processes under pressure. The first is the advantage of using optical methods for the study of high pressure molecular kinetics (see also Attwood and Gutfreund, 1980); the second is the importance of reconstituting membrane components in bilayers of known composition, a point which was also emphasized in section III,B).

Acknowledgement. We are grateful to the Natural Environment Research Council for support; C.J.S. for a Research Studenship and A.G.M. for a project grant.

REFERENCES

Attwood, P.V., Gutfreund, H.(1980) The application of pressure relaxation to the study of the equilibrium between metarhodopsin I and II from bovine retinas. FEBS Letters 119 , 323-326.

Balschi, J.A., Cirillo, V.P., Le Noble, W.J., Pike, M.M., Schreiber, E.C., Simon, S.R., Springer, C.S.(1982) Magnetic resonance studies of metal cation transport across biological membranes: use of paramagnetic lanthanide ions. In : McCarthy CJ, Rhyne JJ and Silber HE (eds) Rare earths in modern science and technology, vol.3, Plenum Press, New York, p.15.

Bourne, P.K., Cossins, A.R.(1982) On the instability of potassium influx in erythrocytes of the Rainbow Trout *(Salmo gairdnerii)* and the role of catecholamine hormones in maintaining *in vivo* influx activity. J. Exp. Biol. 101, 93-104.

Brahms, J.(1977) Temperature dependent changes of chloride transport kinetics in human red blood cells. J. Gen. Physiol. 70, 283-306.

Bruner, L.J., Hall, J.E.(1983) Pressure effects on alamethicin conductance in bilayer membranes. Biophys. J. 44, 39-47.

Cala, P.M.(1983) Volume regulation by red blood cells. Mol. Physiol., 4, 33-52.

Chong, PL.G, Jameson, D.M., Fortes, P.A.G., Weber, G.(1982) Effect of pressure and temperature on $(Na^{+}+K^{+})$-ATPase. Biophys. J. 37, 148a.

Chong, PL.G, Cossins, A.R.(1983) A differential polarized phase fluorometric study of the effect of high hydrostatic pressure upon the fluidity of cellular membranes. Biochem. 22, 409-415.

Chong, PL.G, Cossins, A.R.(1984) Interacting effects of temperature pressure and cholesterol content upon the molecular order of dioleoyl phosphatidylcholine vesicles. Biochem. Biophys. Acta, 772, 197-201.

De Pont J.J.H.H.M., Peters, W.H.M., Bonting, S.L.(1983) Role of cholesterol and other neutral lipids in Na, K-ATPase. In : Hoffman JF , Forbush B.(eds) Current topics in membranes and transport, vol.19, Academic Press,pp.163-166.

Dobler, M.(1981) Ionophores and their structures. Wiley Interscience New York.

Eisenmann, G., Horn, R.(1983) Ion selectivity revisited : The role of kinetic and equilibrium processes in ion perneation through channels. J. Memb. Biol. 76, 197-225.

Ernst, W.G., Morin, J.G.(1982) The environment of the deep sea. Prentice Hall New Jersey, USA.

Goldinger, J.M., Kang, B.S., Chon, Y.E., Paganelli, C.V., Hong, S.K. (1980) Effect of hydrostatic pressure on ion transport and metabolism in human erythrocytes. J. Appl. Physiol. Resp. Environ.Exercice Physiol. 49, 224-231.

Hall, A.C., Ellory, J.C., Klein, R.A.(1982) Pressure and temperature effects on human red cell cation transport. J. Mem. Biol. 68, 47-56.

Heremans, K.(1982) High pressure effects on protein and other biomolecules. Ann. Rev. Biophys. Bioeng. 11, 1-21.

Johnson, S.M., Miller,K.W.(1975) The effect of pressure and the volume of activation of the monovalent cation and glucose permeabilities of liposomes of varying composition. Biochim. Biophys. Acta 375, 286-291.

Kregenow, F.M.(1981) Osmoregulatory salt transporting mechanisms, control of cell volume in anisotonic media. Ann. Rev. Physiol. 43, 493-505.

Lakowics, J.R., Thompson, R.B.(1983) Differential polarized phase fluorometric studies of phospholipid bilayers under high hydrostatic pressure. Biochim. Biophys. Acta 372, 359-371.

Low, P.S., Somero, G.N.(1975) Protein hydration changes during catalysis : a new mechanism of enzymic rate-enhancement and ion activation/inhibition of catalysis. Proc. Nat. Acad. Sci 72, 3305-3309.

Macdonald, A.G.(1967) The effect of high hydrostatic pressure on the cell division and growth of Tetrahymena pyriformis. Expt. Cell Res. 47, 569-580.

Macdonald, A.G.(1984) The effects of pressure on the molecular structure and physiological functions of cell membranes. Phil. Trans. Roy. Soc. Lond B 304, 47-68.

Morild, E.(1981) The theory of pressure effects on enzymes. Adv. Protein Chemistry 34, 93-166.

Nakahara, M.(1972) Ionic solutions under high pressure IV. Effects of pressure on the mobilities and hydration of Bu_4N^+, Me_4N^+, K^+ and Cl^- ions. The Review of Physical Chemistry of Japan 42, 75-84.

Neilson, C.W. (1979) The effect of pressure on anionic hydration. Chemical Physics Letters 68, 247-250.

Nikinmaa, M.(1982) Effects of adrenaline on red cell volume and concentration gradient of protons across the red cell membrane in Rainbow Trout *(Salmo gairdnerii)* Mol. Physiol. 2, 287-297.

Péqueux, A., Gilles, R., Pilwatt, G., Zimmerman, U.(1980) Pressure induced variations of K^+ permeability as related to a possible reversible electrical breakdown in human erythrocytes. Experientia 36, 565-566.

Riddick, D.H., Kregenow, F.M., Orloff, J.(1971) The effect of norepinephrine and dibutyryl cyclic adenosine monophosphate on cation transport in duck erythrocytes. J. Cell. Physiol. 57, 752.

Rona, P.A., Bostrom, K., Laubier, L., Smith, K.L.(1984). Hydrothermal processes at seafloor spreading centres. Plenum publishing corp. New York.

Shelton, C.J., Macdonald, A.G., Péqueux, A., Gilchrist, I.(1984) The ionic composition of plasma and erythrocytes of deep sea fish. J. Comp. Physiol. In Press.

Taniguchi, Y., Yukimoto, K., Suzuki, K., Okahata, U., Kunitake, T. (1982) Pressure effects on model systems of biomembranes-lecithins and dialkyldimethylammonium salts in water. In : Ohtake H, Tamamushi R.(eds) Ions and Molecules in solutions. VI International Symposium of Solute-solvent interactions. Studies in Physical and Theoretical Chemistry 27. pp.457-466.

Taylor,C.D.(1978) The effect of pressure upon the solubility of oxygen in water. Arch. Biochem. Biophys. 191, 375-384.

Trauble, H.(1971) The movement of molecules across lipid membranes: a molecular theory. J.Memb. Biol. 4, 193-208.

Tsuda, M., Govindjee, R., Elrey, T.G.(1983) Effects of pressure and temperature on the M412 intermediate of the bacteriorhodopsin photocycle. Biophys. J. 44, 249-254.

Wann, K.T., Macdonald, A.G.(1980) The effect of pressure on excitable cells. Comp. Biochem. Physiol. 66A , 1-12.

Weber, G., Drickamer, H.G.(1983) The effect of high pressure upon proteins and other biomolecules. Quart. Rev. Biophys. 16,89-112.

PRESSURE EFFECTS ON PHOSPHOLIPIDS AND PROTEINS

K. HEREMANS.

I. INTRODUCTION

In this paper we review recent experimental results and ideas about pressure effects on the behaviour of proteins and lipids. The approach will be physico-chemical. The interested reader is referred to other papers presented in this volume (see also Paton *et al.*,1984) for the biological background to these studies. For the physical chemist, the biological relevance is indeed not the primary motivation for high pressure studies. The results of the physical pressure studies have given us new insights into the behaviour of molecules of biological origin. We will demonstrate this with examples from the lipid and the protein field. Topics not considered here are reviewed elsewhere (Heremans, 1982; Weber and Drickamer, 1983).

II. PHOSPHOLIPIDS

A. Phase transitions

From all the pressure studies effected on biological macromolecules and structures, the results on lipids are the most spectacular. The temperature induced phase transitions of phospholipid water dispersions are shifted by the application of pressure by about 20K/kbar (Heremans, 1982; Macdonald, 1984). This figure corresponds to the change in melting temperature of hydrocarbons with pressure and the relation between both processes can be made quantitatively (Nagle and Wilkinson, 1978). This strong effect has been used as an indicator for the role of lipids in various biological processes where transitions (i.e. effects occurring in a short temperature or pressure range) have been observed (Ceuterick *et al.*, 1978; Thevelein *et al.*, 1979; Tsuda *et al.*, 1983; Chong and Weber, 1983). In a few instances a significantly higher value for dT/dP has been found (50K/kbar, Tsuda *et al.*, 1983). In this respect it is interesting to mention the observations (Yager and Chang, 1983) that pressure destabilizes the non bilayer phase(hexagonal H2) with respect to the lamellar phase by about 40 K/kbar.

The activation volumes for a number of membrane transport processes are quite large (Bruner and Hall, 1983; Macdonald, 1984). These numbers almost certainly include the melting of a few lipid molecules which are involved in the activation process. From the changes in activation volumes for membrane bound enzymes with temperature one can actually calculate that a few lipids are involved in the quasi crystalline region or domain surrounding the enzyme (Ceuterick *et al.*, 1978).

A logical (thermodynamic) consequence of these effects is, that pressure can induce phase transitions in lipids at constant temperature. For the physical chemist temperature and pressure induced transitions need not be the same however. Temperature changes produce both volume and energy effects while pressure produces only volume effects. Thus the kinetic energy as well as the volume available for a molecule will change with temperature at constant pressure. Pressure effects at constant temperature will mostly reduce the available space for the molecules. The strong repulsions between molecules as a consequence of the applied pressure will only occur at much higher pressures than those considered in biological experiments. The different response of a system to temperature and pressure can be observed with techniques that probe the motion of molecules such as NMR and Raman spectroscopy. Here we concentrate on the detection of molecular events in lipid transitions with Raman spectroscopy.

B. Raman spectroscopy

Raman as well as Infra red give information on the vibration of atoms in molecules (Carey, 1982). While IR is an absorption technique, Raman is a scatterring technique. When a light beam is shone upon a sample, light is scattered in all directions due to the interaction of the incoming light with the electrons of the molecules. This is the Rayleigh effect where the scattered light has the same frequency as the incoming light. Due to the coupling between the oscillating electrons and the vibration of the atoms, new scattering intensities araise with a frequency different from the incident light. These frequencies are characteristic for particular bonds between atoms or group of atoms and can therefore be used as an identification of the particular compound (a fingerprint). The intensity of the Raman scattering is about 10^{-6} times less than the Rayleigh scattering. As a consequence, rather high concentrations are needed which may be a serious problem with biological material. In view of the small dimension of the laser beam, only a few microliter are needed however. The principal advantage of Raman over other optical techniques is that no probe molecules are needed which may perturb the phase behaviour of the system. A typical spectrum of a phospholipid shows predominantly the contribution of the hydrocarbon chains. The strongest Raman bands of fully hydrated lipids are observed in the frequency region 2700-3100 cm^{-1} (Fig.1). They are due to the C-H stretching modes of the methylene and methyl groups. The region of 1250-1500 cm^{-1} shows the deformation modes of the methylene groups (Fig.2). Between 1050 and 1150 cm^{-1} we observe the C-C stretching modes.

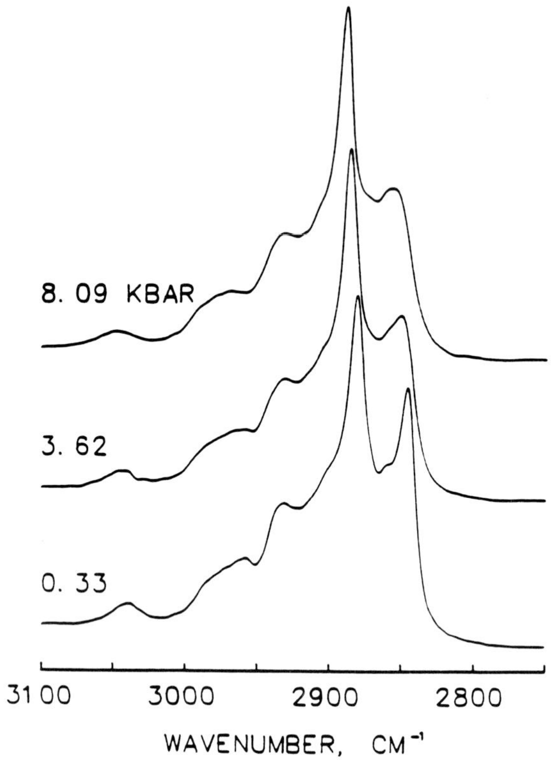

Fig.1. Raman spectra of DPPC aqueous dispersions as a function of pressure in the methylene stretching region (Reproduced with permission from Wong, 1984).

Below 400 cm^{-1} the accordionlike longitudinal acoustic modes of the chains are observed. Bands due the head groups are very much weaker. Here we concentrate on the region of 2700-3100 cm^{-1} and refer to an excellent recent review for an overview of the field (Wong, 1984). The most outstanding peaks are due to the symmetric (2850 cm^{-1})and the asymmetric stretching vibration (2880 cm^{-1}) of the $-CH_2$-groups. In general, the frequencies, the band widths, the intensities and the shape of the bands are very sensitive to interchain dynamics, interactions and packing, phase transitions and intrachain conformation. The bands can therefore be used to follow phase changes in lipids.

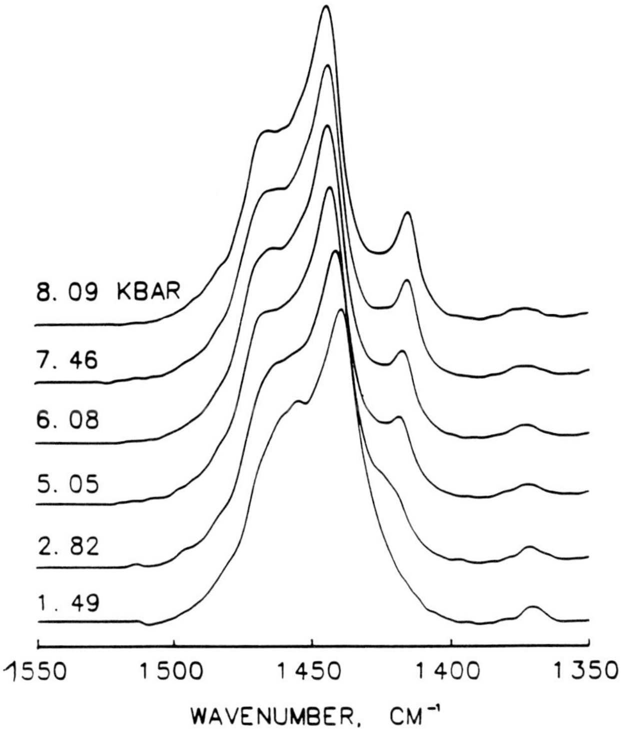

Fig. 2. Raman spectra of DPPC aqueous dispersions as a function of pressure in the methylene scissoring region (Reproduced with permission from Wong, 1984).

Wong *et al.* (1982) studied the effect of pressure on the Raman spectrum of dimyristoyl phosphatidylcholine up to 5 kbar. At 30 °C they observe transitions at 150 bar, at 1 and 2.6 kbar. The changes are correlated with temperature induced transitions at 24°C (liquid crystalline to gel), 14°C (gel to gel pretransition) and a new structural phase change (gel to gel) observed at -60°C.

What happens at these transitions can be inferred from the spectral changes. From the peak intensities it is concluded that interchain interactions increase with increasing pressure. The effect is quite strong in the liquid crystalline phase, less pronounced in the gel phase. From the pressure effect on the widths of the depolarized and weakly polarized bands, the reorientation rate of the chains is seen to be damped with increasing pressure. The bandwidth of the strongly polarized bands (as is the symmetric $-CH_2$-stretching) is an indication for the interchain interaction. In all phases the interchain interaction increases as expected.

A similar picture arises from the analysis of the other Raman bands. We refer to the original paper for a more detailed discussion. The same authors have also made a very detailed study of the phase transitions and structural changes in aqueous sodium oleate. In this particular case a significant difference is observed between the temperature and the pressure effects. For a complete review on this topic we refer the reader to the review by Wong (1984).

While in the previous studies spectra were taken at various pressures after equilibration of the sample, Yager and Peticolas (1982) have initiated the study of the dynamics of the phase transitions with Raman detection. The authors propose a model whereby the rate limiting step in the interconversion between solid and the fluid state is nucleation.

III. PROTEINS

While the effects of pressure on lipids are relatively simple and strong, the effects on proteins are complicated and with a few exceptions , small. We do not intend to review the whole field. This has been done elsewhere (Jaenicke, 1981; Heremans, 1982; Weber and Drickamer, 1983). We concentrate on NMR studies of the reorientational motion of side chain residues in proteins, pressure induced changes in heme proteins and ligand binding to heme proteins.

A. Protein dynamics

In contrast to the classical rigid picture of a protein which emerges from high resolution X-ray crystallography, a number of observations point to a more dynamic nature (Careri *et al.*, 1979). These apparantly conflicting observations can be reconciled by considering the thermodynamic fluctuations of protein molecules (Cooper, 1976). Although we are dealing with macromolecules, proteins are small systems from the thermodynamic point of view. This has consequences when we consider the energy or the volume of a particle. It can be found in standard thermodynamic texts that the mean square energy and volume are given by

$$(dE^2) = KT^2 m\ C_V$$

$$(dV^2) = KT\ V\ B_T$$

where m and V are the mass and the volume of the particle and C_V and B_T the heat capacity at constant volume and the isothermal compressibility. For a typical protein the fluctuations in energy are of the order of 40 kcal/mol and the volume fluctuations of the order of 30 ml/mol. These numbers are quite large compared with energy and volume changes observed for ligand induced conformational changes.

While there is at present no direct evidence for the functional role of the dynamic nature of proteins, there is considerable indirect evidence. This comes from a number of observations from which we review shortly only a few : tryptophan fluorescence quenching, ligand binding, high pressure and compressibility studies.

1. Lakowicz and Weber (1973) have observed the quenching of the fluorescence of internal tryptophan residues in proteins by oxygen. From X-ray diffraction studies these residues were considered inaccessible for the solvent. The quenching is diffusion controlled and the diffusion of oxygen into the protein matrix is considered to be the consequence of the fluctuations in the macromolecular structure.

2. It has been noted since several years that the access to the iron atom for oxygen in myoglobin is not possible judged from the static picture which is obtained from X-ray diffraction. The kinetics of the binding is however fast. It is therefore concluded that protein fluctuations provide transient channels for the ligands to enter the binding pocket. Activation volumes for the binding of oxygen and CO to myoglobin are opposite in sign (Hasinoff, 1974, Frauenfelder

and Wolynes, 1984). This has been interpreted as an indication for the delicate role of the protein in the binding process. Our own high pressure ligand binding studies on myoglobin and cytochrome C give also support to this idea.

3. Wagner (1983) has studied the effect of pressure on the NMR spectrum of bovine trypsin inhibitor. He has estimated the contribution of the fluctuation of the protein structure to the experimental activation volume for the rotation of aromatic side chains. The effect of temperature on the NMR spectrum of cytochrome C has provided a detailed picture of the mobility of the protein structural domains (Moore *et al.*, 1982). The striking difference in thermal stability of the oxidized and reduced form of this protein is reflected in the difference in compressibility of the oxidation states (Eden *et al.*, 1982). As indicated before, the isothermal compressibility is proportional to the volume fluctations.

We now discuss in more detail the dynamical behaviour as inferred from studies of the effect of pressure on the rotational rate (flip rate) of side chain groups in the interior of globular proteins (Wagner, 1980). The rate of rotation of a phenylalanine and tyrosine side chain has been calculated from the high resolution NMR spectra of bovine pancreatic trypsin inhibitor. The rate of flipping decreases with increasing pressure. The activation volume is about 50 $Å^3$ which corresponds to 30 ml/mole. This activation volume can be understood on the basis of a simple physical model. If the aromatic ring is considered to be an ellipsoid, the volume can be calculated to be 82 $Å^3$. The volume of a sphere symbolizing the space occupied by the rotating ring is about 164 $Å^3$. The difference corresponds to a hypothetical activation volume of 82 $Å^3$. The activation volume can also be visualized on the basis of the X-ray structure. It turns out that the atoms surrounding the aromatic rings take about 20 $Å^3$ out of the 164 $Å^3$ needed for the free rotation. The large activation volumes observed are therefore an indication that the actual ring flips occur without collisions in an unoccupied volume. This free volume arises by fluctuations of the protein conformation. This interpretation of the data was given in terms of the transition state theory (Wagner, 1983).

Another interpretation has been given to the same observations by Karplus and McCammon (1981). While there is at present no

experimental way to decide between the two theories it is interesting to consider some of the basic ideas. The theory starts from the assumption that rate phenomena can be dominated by the interaction of the reacting molecules with the solvent environment. In the case of the ring rotation the solvent is the surrounding protein matrix. The obvious difficulty here is to estimate the internal viscosity of the protein. Karplus and MacCammon assume the protein interior to be like liquid hydrocarbons since the aromatic ring is located in an hydrophobic region. While in the previous model no collisions were assumed between the protein atoms and the rings, in this model the pressure dependence of the collision frequency is calculated from the pressure dependence of the viscosity. Compressibility data on the other hand suggest that protein interiors are very compact, solid like. To resolve the observed internal dymamics as observed from NMR and fluorescence data (Lakowics and Weber, 1973), Gavish *et al.* (1983) suggest that proteins are made up from solid like domains which may be associated with secondary structures. It should be clear from this short discussion, that studies of the effect of pressure on various protein related processes may have a major impact on our ideas about proteins.
Recently Eden *et al.* (1982) have observed an increase in apparent compressibility upon oxidation of cytochrome c. The oxidation states of cytochrome c differ by only one electron and crystallographic evidence shows that they have almost identical structures. Several solution properties on the other hand indicate a more flexible structure for the oxidized state. The increase in compressibility indicates that this is reflected in larger volume fluctuations for the oxidized state.

B. Heme proteins

Due to their visible absorption spectrum, the behaviour of heme proteins can easily be studied in an optical cell under pressure. Ogunmola *et al.*(1977) interpreted the spectral changes upon pressurization as structural changes from high spin (open crevice) to the low spin state (closed crevice).

Depending on the number of unpaired electrons in iron, the position of the iron atom is slightly out of the porphyrin plane in the high spin complexes, while it is in the plane in the low spin complexes.

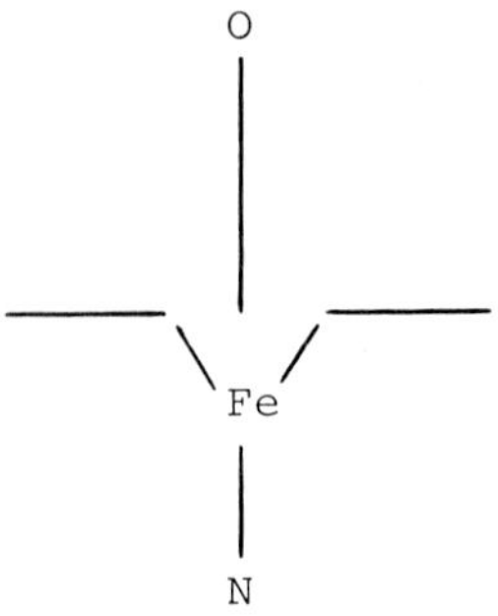

High spin Fe(III)

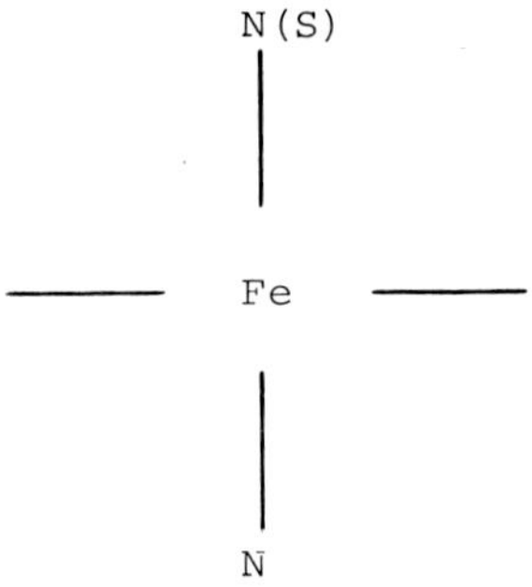

Low spin Fe(III)

From the scheme it is seen that the number of unpaired electrons, and thus the spin state, is controlled by the sixth ligand. The proximal histidine is represented in the lower part of the picture by N. The distal histidine on the side of the sixth ligand is not shown. On the same side, the O indicates the oxygen of the water in metMb. Similarly, N indicates the nitrogen of azide or imidazole. The S represents the sulfur of methionine 80 in cytochrome c. Perutz (1979) has reviewed the influence of the spin state of iron in hemoglobin in relation to the control of oxygen binding.

Morishima *et al.* (1980, 1983) observed pressure induced changes in the proton NMR spectrum of myoglobin , hemoglobin and horse radish peroxidase. Only metMb-N_3^-, metHb-N_3^- and metMb-Im showed striking spectral shifts under pressure, while other complexes such as metMb-CN, metHb-CN and metHb-Im did not show strong changes. This also indicates that the primary effect of pressure is a transformation from high to low spin form. From the NMR results it is concluded that the compression results in a shortening of the ligand iron binding. The observed volume changes associated with the high to low spin conversion for metMb-N_3^- suggest that also protein conformation changes are involved. This was studied in more detail (Morishima and Hara, 1983). First it was found that the distal Histidyl NH resonance (the side of the heme group where the oxygen binds), is more affected by pressure than the proximal Histidine. Secondly, sperm whale MbCN is more stable than horse MbCN against pressure at least in the heme vicinity where the study was located. At the heme, the only difference is at position 67 where a Threonine is changed to Valine in horse Mb. These results are very interesting in view of the suggestion of Gavish *et al.* (1983) that proteins are build up from solid like domains. Perhaps the heme crevice may then be considered as an interface between suchs domains.

C. Ligand binding to heme proteins

1. Myoglobin

In the preceding sections molecular interactions in proteins were studied by NMR spectroscopy. Together with fluorescence and Raman studies these techniques probe the dynamics of part of the protein molecules. A different approach is to look for the response of the protein molecule towards the binding of small molecules.

A number of studies have been devoted to the effects of pressure on the binding of oxygen and CO to heme, Mb and Hb. For the binding of CO an activation volume of + 2 ml and - 9 ml was found for heme and Mb respectively (Hasinoff, 1974). For the binding of oxygen to Mb an activation volume of + 8 ml and a reaction volume of - 3 ml was found. While the negative reaction volume may be explained by the change in spin state without the involvement of a change of the protein, the activation volume points to a participation of the protein. From X ray studies it is known that oxygen is to large to enter the heme pocket in Mb (Perutz, 1979). This would make a change in the protein (an opening of the heme crevice) a likely mechanism. Recently a molecular door stop was observed for the binding of phenylhydrazine (Ringe *et al.*, 1984). The opposite activation volume observed for CO binding, indicates however that the same protein controls the incoming of a similar ligand in a rather different way. In our own high pressure stopped flow and binding studies we obtained for the binding of Imidazol and azide results which point to the same mechanism (Table 1).

TABLE 1. Activation and reaction volumes for the binding of ligands to sperm whale Mb (298K)

	ΔV	$\Delta V^{\#}$	Ref
Mb + O_2	- 3 ml	+ 8 ml	(a)
metMb + Im	0 ml	+ 3 ml	(b)
metMb + N_3^-	- 9 ml	+14 ml	(b)

(a) Hasinoff (1974) (b) Bormans and Heremans (Unpublished data)

The results in Table I refer to Mb for the binding of oxygen and to metMb for the other ligands. It is not known what influence the redox state of the iron has on the conformation of the protein. For cytochrome c the change in redox state has very strong implications on the stability of the protein although the conformation as obtained from X ray diffraction, is not drastically different (vide infra).

In the reactions from Table I the iron atoms in the proteins do not have the same spin configuration. As already noted Fe(II)Mb is a high spin compound while the oxygen complex is a low spin compound. When we interpret the volume changes for the ligand binding we have to take into account the pressure dependence of the spin configuration of the protein compounds. If we concentrate on the Fe(III)Mb or metMb we have to consider that the unliganded and liganded state of the protein have a different spin configuration which are a thermal mixture of high spin (hs) and low spin (ls). We consider the following reaction scheme :

$$\begin{array}{ccc} \text{metMb(H}_2\text{O)hs} + \text{L} & \rightleftharpoons & \text{metMb(L)hs} \\ \updownarrow & & \updownarrow \\ \text{metMb(H}_2\text{O)ls} + \text{L} & \rightleftharpoons & \text{metMb(L)ls} \end{array}$$

we define the following equilibrium constants :

$$K = \text{metMb(H}_2\text{O)ls} \,/\, \text{metMb(H}_2\text{O)hs},$$

$$K(L) = \text{metMb(L)ls} \,/\, \text{metMb(L)hs}$$

These expressions are for the spin equilibria of the unliganded and liganded forms of the protein. For the binding we consider the possibility that the ligand can bind to the high spin or the low spin form.

$$K(hL) = (\text{metMb(L)hs}) \,/\, (\text{metMb(H}_2\text{O)hs})\ (L)$$

and

$$K(lL) = (\text{metMb(L)ls}) \,/\, (\text{metMb(H}_2\text{O)ls})\ (L).$$

The following relation exists between these equilibrium constants :

$$K \cdot K(lL) = K(L) \cdot K(hL)$$

The experimentally observed equilibrium constant K(obs) is a composite quantity of these constants :

$$K(obs) = K(hL)\ \{(1 + K(L)) / (1 + K)\}$$

In order to find the volume changes for the ligand binding, we need the volume changes for the spin equilibria. The results are summarized, together with values for the equilibrium constants, enthalpy and entropy changes in Table II.

Table II. Thermodynamic data for spin equilibria of metMb(K=hs/ls)(a)

	K	ΔH	ΔS	ΔV(b)
Mb(H_2O)	8.09	0.06	4.4	3
Mb (OH)	1.85	1.23	5.8	5
Mb(N_3^-)	0.2	3.8	9.6	9
Mb(Im)	0.16	10.8	32	12

(a) Sano *et al.* (1981)

(b) Bormans and Heremans (Unpublished data).

It is interesting to note that the volume changes are correlated with the entropy changes. A similar trend has been observed for inorganic Fe(II) complexes (McGarvey *et al.*, 1984) as a function of solvent composition. The role of the solvent is taken over by the protein and the ligand. It is at present not clear whether geometric factors are important, but experiments in progress with horse Mb indicate that this may be the case.

We now turn our attention to the volume changes for the ligand binding as reported in Table I. The small volume changes observed by Hasinoff for the binding of oxygen to Mb can be attributed to spin changes but in the absence of data on the volume changes for spin equilibria of deoxyMb and oxyMb this conclusion is tentative.

For our own data on the binding of azide and imidazole we can be more specific. From the relation between the equilibrium constants we obtain the following expression for the volume changes :

$$\Delta V(lL) + \Delta V(H_2O) = \Delta V(hL) + \Delta V(L)$$

We can now appreciate that if differences are observed for the spin transitions, other thermodynamic data will be different for the same ligand to either the low or the high spin configuration of the protein. For azide the volume changes are negative while for imidazole the sign is different. The sign for the binding to the hs configuration is positive, for the ls configuration it is negative. From the negative volume changes for the binding of azide, we conclude that no electrostatic factors are involved. The formation of a bond would be sufficient to account for the negative sign, but geometric factors may also explain the difference between the hs and the ls configuration. For the binding of imidazole we invoke essentially the same mechanism, keeping in mind that this ligand is larger and bound in a different way (perpendicular to the heme plane in the case of imidazole and bend in the case of azide). X ray diffraction of the Im complex with metMb shows that the distal histidine changes position with the ligand in the pocket (Bologonesi *et al.*, 1982). A general model has recently been proposed for the binding of anions to high spin ferric hemoproteins (Jain and Kassner, 1984).

The activation volumes indicate the response of the same protein to the different ligands. Similar observations were made in high pressure studies performed at low temperatures for the binding of CO and O_2 to sperm whale and horse Mb. See Table III

Table III. Activation volumes for the binding of ligands to metMb (160 K)(a)

	CO	O_2
Sperm whale Mb	- 16 ml	+ 10 ml
horse Mb	- 9 ml	- 3 ml

(a) Sorensen (1980)

As indicated before, the only difference in the heme pocket between horse and sperm whale Mb, is the substitution of one amino acid. The difference in activation volume point to the subtle control of the protein on the binding of the ligand. A more quantitative explanation can at present not be given (Frauenfelder and Wolynes, (Unpublished data).

2. Cytochrome c

Another approach to study the influence of the protein to ligand binding is to take a protein where the sixth heme ligand is already occupied by a protein ligand. In cytochrome c this is a sulfur atom from methionine 80. The Fe-S bond can be broken e.g. by a change to alkaline pH, but it is not certain which ligand replaces the sulfur atom (Bosshard,1981). The iron-sulfur bond is also broken when an external ligand like Im or azide binds (Schejter and George, 1964). Table IV shows the results of our pressure studies.

Table IV. Activation and reaction volumes for the binding of ligands to cytochrome c(a)

	ΔV	$\Delta V^{\#}$
ferricyt + Im	+ 7 ml	+ 20 ml
ferricyt + N_3^-	- 13 ml	+ 20 ml

(a) Bormans and Heremans (Unpublished data).

In contrast to the ligand binding to Mb, no changes in spin state occurs, the iron being in the low spin state.

Here we relate the reaction volume difference between Im and N_3^- to the different geometrical positions of these ligands as explained for Mb. It is also of interest to note that for Mb and cyt c the reaction volume is more positive for Im than for N_3^-. The similar activation volume for both ligands reflects in our opinion a similar change in the protein conformation in contrast to the situation for Mb.

While it seems that we have a qualitative picture of the ligand binding in terms of volume changes, a more quantitative picture would be obviously desirable. If the oxygen affinity to Hb could be regulated by simple mechanical devices on a molecular level (Perutz, 1979), then we have here a few examples showing how this is reflected in terms of volume changes.

The experimental observation that the activation volume parameters for Mb are dependent on the geometry of the ligand binding are in our opinion a strong case in favour of the free volume model (Wagner, 1980). For the collision model (Karplus and McCammon, 1981) one would expect similar activation parameters for different ligands as the theory only takes into account the changes in internal viscosity of the protein. As indicated, for cytochrome c, the results point towards a different mechanism.

IV. PROTEIN DENATURATION

The processes discussed in the previous sections are phenomena which may be considered to be localized in the protein. Protein denaturation on the other hand is thought to be a change from the globular to the coiled state. As discussed elsewhere (Heremans, 1982; Weber and Drickamer, 1983), the various techniques employed to study this denaturation make use of probe molecules and might therefore reflect only localized changes. Information on the changes in the hydrodynamic radius of proteins can be obtained from dynamic light scattering. Temperature studies have revealed that lysozyme and chymotrypsinogen increase their radius by about 10%. Lysozyme has now been studied under pressure. The increase in radius is about 10% (Nystrom and Roots, 1984). Fluorescence polarization measurements yield information on the overall rotational rate of a protein. Probe molecules have to be introduced however in order to circumvent the heterogeneous emission arising from different tryptophan residues. For lysozyme a 60% increase in apparent volume is observed. For chymotrypsinogen the results are time dependent and only partially reversible (Chrysomallis *et al.*, 1981). Techniques which have not previously been used to study protein denaturation under pressure such as NMR and Raman spectroscopy are expected to give important information on the denatured state of a protein.

A high pressure Raman study has recently been made on the denaturation of lysozyme (Heremans and Wong, unpublished observations). Only minor changes occur up to 3.6 kbar. At 5.5 kbar, irreversible denaturation and precipitation occurs. The pressure at which this denaturation occurs corresponds to changes observed in the fluorescence polarization studies (Chrysomallis *et al.*, 1981). The Raman observations also support the interpretation of the dynamic light scattering measurements which indicate increasing intermolecular interactions with increasing pressure (Nystrom and Roots, 1984). It remains to be investigated whether other proteins show the same behaviour.

V. CONCLUSION

The present review only a number of selected high pressure studies in the protein and lipid field indicates the complex behaviour of these systems. Especially with regard to proteins, one does indeed not expect simple answers from a study of a complex system. Also, the models we use to describe the molecular details of these processes are macroscopic. It is hoped that spectroscopic techniques such as fluorescence polarization, Raman and NMR spectroscopy, will be our guides towards molecular models. The Raman spectroscopy work on phospholipids (Wong, 1984) and the NMR work of Morishima *et al.* (1983) on heme proteins, are pioneering in this respect. The high pressure stopped-flow technique (Heremans *et al.*, 1980) has now been adapted to measurements at low temperatures (Balny *et al.*, 1984) with the possibilities to stabilize enzyme intermediates and study the effect of pressure on their formation.

The combination of lipids and proteins into the highly organized biological membranes creates an even more challenging medium for analysis. While pressure studies have contributed to our understanding on protein lipid interactions (Heremans *et al.*, 1982), more complex phenomena are more elusive. Suffice it to refer to the recent review by Macdonald (1984) on the effect of pressure on various passive and active membrane transport phenomena. But here again we expect important contributions from high pressure studies to the understanding of these fundamental processes.

REFERENCES

Balny, C., Saldana, J.L., Dahan, N.(1984) High pressure stopped flow spectrometry at low temperatures. Anal. Biochem. 139, 178-190.

Bolognesi, M., Cannillo, E., Ascenzi, P., Giacometti, G.M., Merli, A. Brunori, M.(1982) Reactivity of ferric aplasia and sperm whale myoglobins towards imidazole. X-ray and binding study. J. Mol. Biol. 158, 305-315.

Bosshard, H.R.(1981) Alkaline isomerization of ferricytochrome C J. Mol. Biol. 153, 1125-1149.

Bruner, L.J., Hall, J.E.(1983) Pressure effects on alamethicin conductance in bilayer membranes. Biophys. J. 44, 39-47.

Careri, G., Fasella, P., Gratton, E.(1979) Enzyme dynamics : the statistical physics approach. Ann. Rev. Biophys. Bioeng., 8, 69-97.

Carey, P.R. (1982) Biochemical applications of Raman and Resonance Raman spectroscopy. Academic Press, New York, London.

Ceuterick, F., Peeters, J., Heremans, K., De Smedt, H., Olbrechts, H. (1978) Effect of pressure, detergents and phospholipase on the break in the Arrhenius plot of *Azotobacter* nitrogenase. Eur. J. Biochem. 87, 401-407.

Chong, P.L.G., Weber, G.(1983) Pressure dependence of DPH fluorescence in single component phosphatidylcholine liposomes. Biochemistry 22, 5544-5550.

Chryssomallis, G.S., Torgerson, P.M., Drickamer, H.G., Weber, G. (1981) Effect of hydrostatic pressure on Lysozyme and Chymotrypsinogen detected by fluorescence polarization. Biochemistry 20, 3955-3959.

Cooper, A.(1976) Thermodynamic fluctuations in protein molecules. Proc. Nat. Acad. Sci. (USA) 73, 2740-2741.

Eden, D., Matthew, J.B., Rosa, J.J., Richards, F.M.(1982) Increase in apparent compressibility of cytochrome c upon oxidation. Proc. Nat. Acad. Sci. 79, 815-819.

Gavish, B., Gratton, E., Hardy, C.J.(1983) Adiabatic compressibility of globular proteins. Proc. Nat. Acad. Sci. 80, 750-754.

Hasinoff, B.B. (1974) Kinetic activation volumes of the binding of oxygen and carbon monoxide to hemoglobin and myoglobin studied on a high pressure laser flash photolysis apparatus. Biochemistry 13, 3111-3117.

Heremans, K., Snauwaert, J., Rijkenberg, J.(1981) Stopped flow apparatus for the study of fast reactions in solution under pressure. Rev. Sci. Instrum. 51, 806-808.

Heremans, K.(1982) High pressure effects on proteins and other biomolecules. Ann. Rev. Biophys. Bioeng. 11, 1-21.

Heremans, K., De Smedt, H., Wuytack, F.(1982) Pressure effects on protein lipid interactions. Biophys. J. 37, 74-75.

Jaenicke, R.(1981) Enzymes under extremes of physical conditions. Ann. Rev. Biophys. Bioeng. 10, 1-67.

Jain, A., Kassner, R.J.(1984) Cyanate binding to the ferric heme octapeptide from cytochrome c : A model for anion binding to high spin ferric hemoproteins. J. Biol. Chem. 259, 10309-10314.

Karplus, M., McCammon, J.A.(1981) Pressure dependence of aromatic ring rotation in proteins : a collisional interpretation. FEBS Letters 131, 34-36.

Lakowicz, J.R., Weber, G.(1973) Quenching of protein fluorescence by oxygen. Detection of structural fluctuations in proteins on the nanosecond time scale. Biochemistry 12, 4171-4179.

Macdonald, A.G.(1984) The effects of pressure on the molecular structure and physiological functions of cell membranes. Phil. Trans. R.Soc. Lond. B 304, 47-68.

McGarvey, J.J., Lawthers, I, Heremans K., Toftlund, H.(1984). Spin relaxation dynamics in Iron (II) complexes : Solvent effects on the activation and reaction volumes. J. Chem. Soc. Chem. Commun. 1575-1576.

Moore, G.R., Huang, Z., Eley, C.G.S., Barker, H.A., Williams, G., Robinson, M.N., Willimans, R.J.P.(1982) Electron transfer in Biology : The function of Cytochrome c; Faraday Discuss Chem. Soc. 74 , 311-329.

Morishima, I., Ogawa, S. and Yamada, H.(1980) High pressure H NMR studies of hemoproteins. Pressure induced structural change in heme environments of myoglobin,hemoglobin and horseradish peroxidase. Biochemistry 19, 1569-1575.

Morishima, I., Hara, M.(1983) High pressure NMR studies of hemoproteins. Pressure induced structural changes in the heme environments of ferric low spin metmyoglobin complexes. Biochemistry, 22, 4102-4107.

Nagle, J.F., Wilkinson, D.A.(1978) Lecithin bilayers. Density measurements and molecular interactions. Biophys. J. 23, 159-175.

Nystrom, B., Roots, J.(1984) Quasielastic light scattering studies of the pressure induced denaturation of lysozyme. Makromol. Chem. 185, 1441-1447.

Ogunmola, G.B., Zipp, A., Chen, F., Kauzmann, W.(1977) Effects of pressure on visible spectra of complexes of myoglobin, hemoglobin, cytochrome c and horse radish peroxidase. Proc. Nat. Acad. Sci. 74, 1-4.

Paton, W., Elliott, D.H., Smith E.B. (1974). Diving and life at high pressures. Phil. Trans. R. Soc. Lond. B304, 1-197.

Perutz, M.F.(1979) Regulation of oxygen affinity of hemoglobin. Ann. Rev. Biochem. 48, 327-386.

Ringe, D., Petska, G.A., Kerr, D.E., Ortiz de Montellano, P.R.(1984) Reaction of myoglobin with phenylhydrazine : a molecular doorstop. Biochemistry, 23, 2-4.

Sano, T., Ishibashi, J., Ikeda, N., Yasunaga, T., Ogawa, S., Morishima, I.(1981) Kinetic studies of spin interconversions in ferric mixed-spin derivatives of myoglobin. Biopolymers 20, 187-199.

Schejter, A., George, P.(1964) The 695 nm band of ferricytochrome c and its relationship to protein conformation. Biochemistry, 3, 1045-1049.

Sorensen, L.B.(1980) The influence of pressure on the low temperature kinetics of myoglobin. PhD Thesis, University of Illinois.

Thevelein, J., Van Assche, J.A., Carlier, A., Heremans,K.(1979). Heat activation of spores : Thermodynamics and effect of alcohols, furfural and high pressure. J. Bacteriology 139, 478-485.

Tsuda, M., Govindjee, R., Ebrey, T.G.(1983). Effects of pressure and temperature on the M412 intermediate of the bacteriorhodopsin photocycle. Biophys. J. 44, 249-254.

Wagner, G.(1980) Activation volumes for the rotational motion of interior aromatic rings in globular proteins determined by high resolution H NMR at variable pressure. FEBS Letters 112, 280-284.

Wagner, G.(1983) Characterization of the distribution of internal motions in the basic pancreatic trypsin inhibitor using a large number of internal NMR probes. Quart. Rev. Biophys. 16, 1-57.

Weber, G., Drickamer, H.G.(1983) The effect of high pressure upon proteins and other biomolecules. Quart. Rev. Biophys. 16, 89-112.

Wong, P.T.T.(1984) Raman spectroscopy of thermotropic and high pressure phases of aqueous phospholipid dispersions. Ann. Rev. Biophys. Bioeng. 13, 1-24.

Wong, P.T.T., Murphy, W.F., Mantsch, H.H.(1982) Pressure effects on the Raman spectra of phospholipid membranes. J. Chem. Phys. 76, 5230-5237.

Yager, P., Chang, E.L.(1983) Destabilization of a lipid non bilayer phase by high pressure. Biochem. Biophys. Acta 731, 491-494.

Yager, P., Peticolas, W.L.(1982) The kinetics of the main phase transition of aqueous dispersions of phospholipids induced by pressure jump and monitored by Raman Spectroscopy. Biochim. Biophys. Acta 688, 775-785.

B EFFECTS OF HIGH PRESSURE ON NEUROPHYSIOLOGICAL PROCESSES

ELECTRICAL AND MECHANICAL FUNCTIONS OF HEART CELLS AT HIGH HYDROSTATIC PRESSURE

P.M. HOGAN.

I. INTRODUCTION

Prerequisite to safe, productive sojourns to the oceans' depths is a thorough understanding of the biological effects of hydrostatic pressure, the *sine qua non* of all hyperbaric environments. Of immediate and practical concern are those actions of elevated hydrostatic pressure that alter normal function in vital tissues of the body. In this chapter I will discuss experimental findings that show hydrostatic pressure to be a factor determining the rate and rhythm of cardiac excitation and the force of cardiac contraction.

II. HYDROSTATIC PRESSURE EFFECTS ON CARDIAC EXCITATION

Cardiac muscle has the unique property of initiating its own excitatory impulse and distributing that impulse in a time orderly fashion that maximizes the mechanical efficiency of the heart during each cardiac cycle. Electrophysiological studies have provided a coherent description of cardiac activation in terms of ionic events occurring across the membranes of individual cardiac cells. It is from this perspective that we have examined the effects of elevated hydrostatic pressure on excitability, conduction, refractoriness, and automaticity, those phenomena that in the aggregate are the basis for cardiac excitation.

The common feature of all excitable cells is an electrically polarized cell membrane. In heart cells the steady-state resting potential of -85 to -100 mV is due to a large outward potassium flux relative to a small inward sodium flux. Under conditions of rapid stimulation electrogenic extrusion of sodium from the cell also contributes to resting potential. Microelectrode measurements of resting potential in heart cells (Doubt and Hogan, 1979) show that elevated pressure causes generalized depolarization and reduces the hyperpolarization caused by rapid electrical stimulation. These effects occur immediately with a change in pressure, are reversible, and are exacerbated by low temperature (table 1). Johnson and Miller (1975) found that pressure decreases both passive potassium permeability and valinomycin-mediated potassium permeability in artificial lipid bilayers. Several studies have demonstrated reduced passive flux in both excitable and non-excitable cells (for a review see Wann and MacDonald, 1980). It is plausible then that pressure-induced depolarization of heart cells is due in part to reduced potassium permeability. The reduction in "overdrive" hyperpolarization on the other hand is most likely related to pressure inhibition of the electrogenic sodium-potassium pump. This interpretation is consistent with the finding of Goldinger *et al.*(1980) who have direct evidence for pump inhibition in the human erythrocyte at pressure. As a further test of the pump inhibition hypothesis we have "short circuited" the electrogenic pump by raising extracellular potassium and have observed that pressure no longer causes depolarization.

TABLE 1

Effects of hydrostatic pressure, temperature and beating frequency on resting potential in heart cells

	Stimulation Frequency (Hz)		Temperature (°C)	
	1	4	37	27
1 ATA	-96.7 ± 1.3	-100.3 ± 1.4	-99.3 ± 2.3	-94.6 ± 1.3
150 ATA	-92.3 ± 1.0	-94.7 ± 1.4	-94.2 ± 1.5	-83.6 ± 2.8

Generalized depolarization may be expected to increase cell excitability but we have observed that excitability is in fact reduced by pressure (Doubt and Hogan, 1978). The stimulus strength-duration curves shown in Fig.1 illustrate this point. At 150 ATA the charge requirement for activation is increased significantly above the 1 ATA control level. Fitting the Hill-Lapique equation to these data reveals a marked increase in the membrane time constant without a change in rheobasic current strength. The implication of these findings is that pressure reduces excitability by increasing membrane capacitance. Substantiation of this possibility requires more direct measurements of membrane capacity.

Regardless of the underlying mechanism it is clear that the decrease in excitability seen at pressure has a significant effect on action potential propagation. Correlated with the decrease in excitability is a marked slowing of the conducted impulse (Doubt and Hogan, 1978). The graphical relation shown in Fig.2 is qualitatively typical of all tissues studied, and shows that the time required for the impulse to propagate between two fixed sites increases proportionally with pressure. In this example conduction time had increased by 40 percent at 150 ATA. It is noteworthy to point out that the application of 2.5-5 atmospheres of nitrous oxide reversed the deleterious effects of pressure on both excitability and conduction.

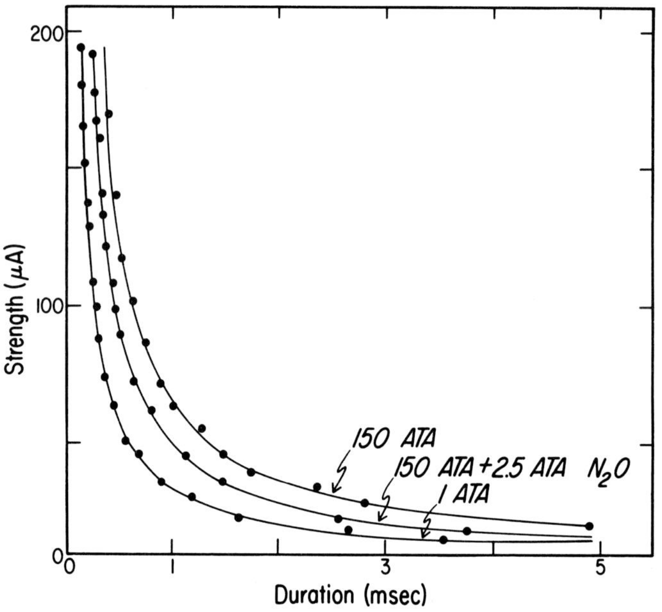

Fig.1. *Pressure effects on excitability in rabbit atria - N_2O antagonism : with an increase in hydrostatic pressure the stimulus strength-duration curve shifted upward and to the right denoting a decrease in excitability. This effect was partially reversed by the addition of 2.5 ATA N_2O. Reproduced by permission from Doubt and Hogan, 1978.*

Other factors must be considered in the response of myocardial conduction to elevated pressure. Under normal circumstances the energy required to drive propagation is derived from the spatial voltage gradient generated during the rapid depolarization phase of the action potential. Furthermore, the upstroke of the cardiac action potential, like that of the nerve axon, contains information about the time and voltage dependent characteristics of participating sodium channels and thus provides a means for examining the influence of pressure on the genesis of excitatory sodium current.

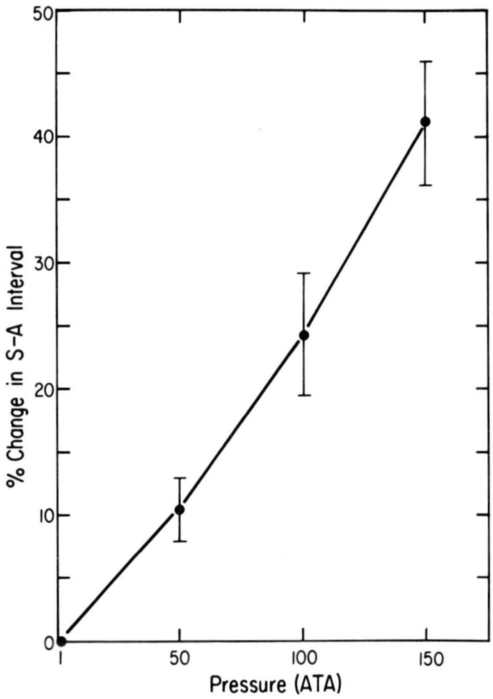

Fig.2. Percentage change in stimulus to atrial (S-A) response interval in the rabbit heart as a function of increasing hydrostatic pressure. An increase in the time required for an impulse to conduct between two fixed sites indicates a decrease in conduction velocity. Each point is the mean ± SE of 12 experiments. Reproduced by permission from Doubt and Hogan, 1978.

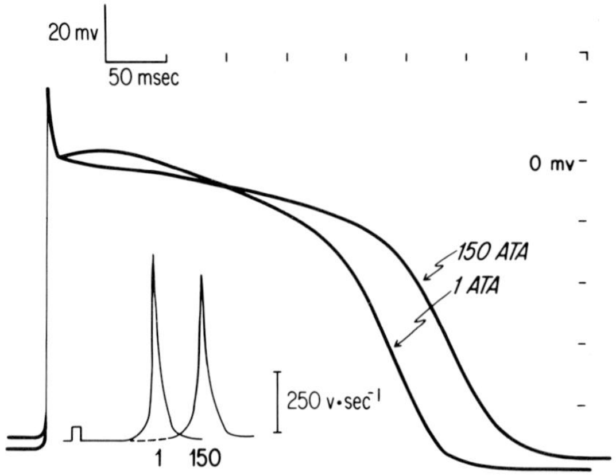

Fig.3. *Superimposed action potential tracings from a single Purkinje fiber from the dog ventricle at 1 and 150 ATA. Also shown is the electronically differentiated upstroke of each action potential displayed at a rapid sweep speed. The time interval between differentiator tracings represents a slowing of conduction at 150 ATA. Reproduced by permission from Doubt and Hogan, 1979.*

In heart cells, as pressure increases, maximum upstroke velocity (V_{max}) decreases with a concomitant slowing of conduction as shown in Fig.3.

Quantitative information regarding the effect of pressure on the "fast sodium system" is provided by the relation between V_{max} and "take-off potential", i.e. the potential from which the action potential is initiated. The predominent downward shift in this relation, as demonstrated in Fig.4, indicates that pressure reduces the availability of excitatory sodium current over the full range of possible resting potentials, a result that contributes directly to pressure-induced slowing of impulse propagation. It is of interest to note that this decrease in sodium availability is similar to that observed in the hydrostatically compressed squid axon (Conti *et al.*, 1982) where slow sodium activation kinetics and increased steady-state sodium inactivation at pressure have been demonstrated directly.

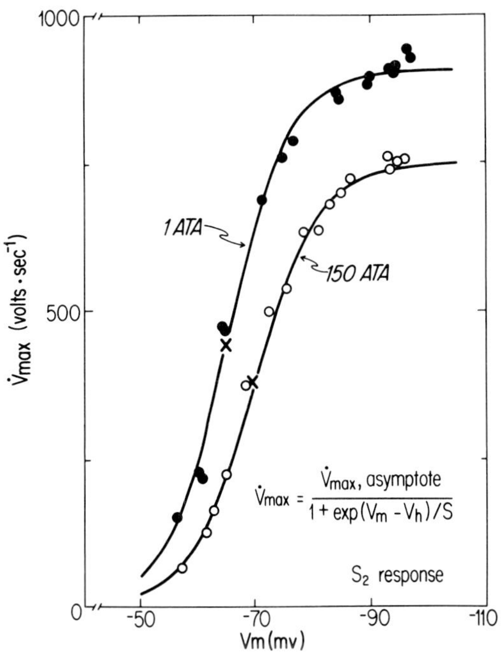

Fig.4. Effect of hydrostatic pressure on the relationship between maximum upstroke velocity (V_{max}) of a canine cardiac Purkinje fiber and the membrane potential at which it was initiated (V_m). Note that increased pressure depresses this curve suggesting that less sodium conductance is available for rapid depolarization. A premature stimulus protocol was used to generate the data. Reproduced by permission from Doubt and Hogan, 1979.

In this regard the apparent activation volume of 32 $cm^3.mol^{-1}$ calculated for sodium activation in the squid axon compares favorably with the activation volume of 37.7 $cm^3.mol^{-1}$ calculated for maximum upstroke velocity in ventricular Purkinje fibers.

Another important effect of pressure on the cardiac action potential is also illustrated in Fig.3, namely that an increase in pressure slows repolarization and increases action potential duration (Doubt and Hogan, 1979). Fig.5 shows this effect over a wide range of diastolic intervals.

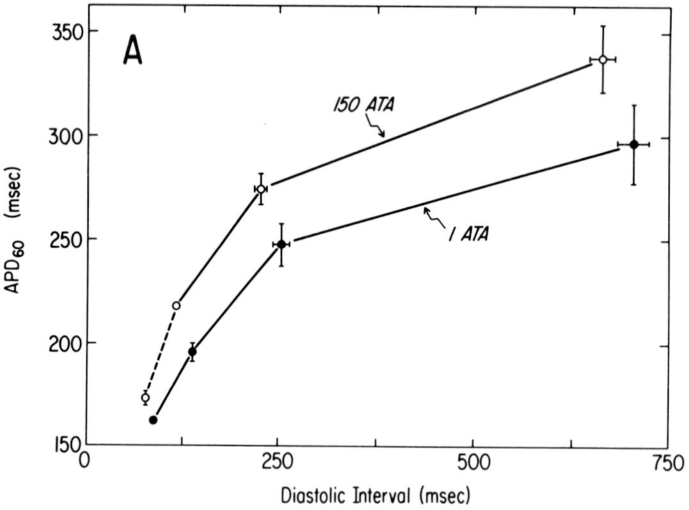

Fig.5. Effect of hydrostatic pressure on the action potential duration-diastolic interval relation in canine cardiac Purkinje fibers. Note that increased pressure increased action potential duration over the range of diastolic intervals tested. Reproduced by permission form Doubt and Hogan, 1979.

The rather uniform action of pressure to lengthen the action potential over this range suggests a reduction in a time-independent outward current similar to that proposed for pressure-induced depolarizations, thus providing further circumstantial evidence for pressure inhibition of electrogenic sodium extrusion in heart cells. The functional consequence of prolonged action potential duration is illustrated in Fig.6, where the combined effect of prolonged duration and decreased sodium availability resulted in an increase in the refractory period of the cell. The extrasystolic response shown in Fig.6 was markedly attenuated at 150 ATA and propagated at a rate sufficiently slow to contribute to the production of a reentrant arrhythmia as will be discussed later in this chapter.

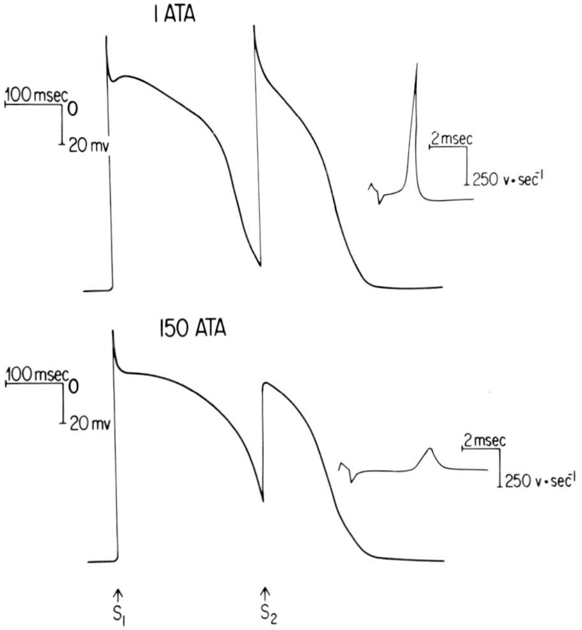

Fig.6. Canine cardiac Purkinje action potentials showing response to premature S_2 stimulation at 1 and 150 ATA. Differentiated upstrokes of S_2 responses are shown to the right of each tracing. At 150 ATA the S_2 response was attenuated and conducted at a much reduced rate. Reproduced by permission from Doubt and Hogan, 1979.

Repolarization in cardiac pacemaker cells is followed by slow diastolic depolarization of membrane potential to the point of initiation of the next excitatory impulse. Elevated hydrostatic pressure slows the rate of diastolic depolarization and thus slows the beating frequency of the heart (Fig.7), a response that we have called hyperbaric bradycardia (Ornhagen and Hogan, 1977).

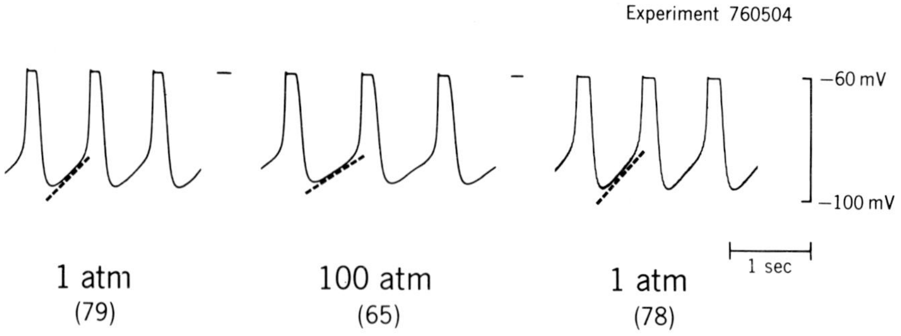

Fig.7. *Effect of increased hydrostatic pressure on diastolic membrane potential in canine cardiac Purkinje fiber. Numbers in parentheses indicate spontaneous beating frequency of preparation in beats.min^{-1}. Dashed lines indicate maximum slopes of diastolic depolarization. Note correspondence between pressure-induced slowing of diastolic depolarization. Reproduced by permission from Ornhagen and Hogan, 1977.*

Calculations based on the linear additive model of Brauer *et al.* (1982) indicate that the relative intrinsic potency of hydrostatic pressure to elicit bradycardia is comparable to that for pressure convulsions but is substantially larger than that for pressure reversal of anesthesia. Measurable antagonism to the effect of pressure on beating frequency has been demonstrated for several inert gases (Ornhagen, 1979). The ascending order of intrinsic potency of these gases to reverse hyperbaric bradycardia is helium $<$ neon $<$ nitrogen $<$ nitrous oxide. Nitrous oxide has also been used, as mentioned earlier , to demonstrate the reversibility of pressure-induced changes in cardiac excitability and conduction.

It should be apparent from the foregoing discussion that pressure exerts an ubiquitous influence over the electrical properties of the heart. The phenomena considered represent complex membrane events which involve macromolecular reactions that proceed with pressure sensitive volume changes. At this point in our knowledge it is not possible to identify and characterize the specific rate limiting steps governing those events but it is, nevertheless, of practical value to catalog the actions of pressure according to their relative potency to produce changes in the relevant end-point phenomena. Calculation of apparent activation volume based on phenomenological rate constants provides a means for making this comparison.

TABLE 2

Apparent activation volume, $\Delta V^{\ddagger}$, for cardiac electromechanical events

Event	Tissue	$\Delta V^{\ddagger}$ ($cm^3.mol^{-1}$)
Repolarization	Dog Purkinje Fiber	24
Upstroke Velocity	Dog Purkinje Fiber	38
Conduction	Dog Purkinje Fiber	37
	Rabbit Atrium	58
Excitability	Rabbit Atrium	82
Beating Frequency	Mouse Sinus Node	93
Pacemaker Potential	Dog Purkinje Fiber	105
Twitch Tension	Cat Papillary Muscle	-200

Table 2 shows apparent activation volumes arranged in order of ascending potency and indicates that refractoriness is least affected, conduction is intermediate and automaticity is most sensitive to changes in hydrostatic pressure. Further, it is noteworthy to point out that the apparent activation volumes for electrical phenomena are all positive, indicating that the predominant effect of pressure is inhibitory on underlying processes. Correspondence between the apparent activation volumes for the rapid upstroke and conduction in dog Purkinje fibers is consistent with a cause and effect relationship between these variables. Close correspondence between the rate of diastolic depolarization (pacemaker potential) and beating frequency, suggest a similar relationship for these variables.

In view of the fact that pressure tends to depress electrical activity in the heart it is important to consider the potentially deleterious effects of pressure on cardiac rate and rhythm. Cardiac arrhythmia is defined as a disturbance in normal impulse initiation and/or conduction. By this definition elevated hydrostatic pressure is inherently arrhythmogenic, for as we have seen, elevated pressure causes bradycardia and depressed conduction throughout the myocardium. Given the magnitude of these effects, however , is it reasonable to suggest that the action of pressure on cardiac excitation may pose a real threat to man working under the sea ? With regard to hyperbaric bradycardia the answer is probably "no". The direct effect of pressure to slow the heart is more than likely offset by the many factors that normally control heart rate with little if any compromise in the pumping action of the heart. Conduction disturbances on the other hand are far more serious. Slow conduction is a prerequisite for the most severe and life threatening of the cardiac arrhythmias.

TABLE 3

Arrhythmias encountered during hydrostatic compression

	Tissue	Pressure (ATA)	Temp. (^{o}C)	Rate (pulses.min^{-1})	Arrhythmia
I	Sinus Node*	60-150	23-32		Bradycardia, Exit Block
II	Atria (2 of 8)	100-150	30	150-200	3:2 Block, 2:1 Block
III	Purkinje	100	37	60	Coupled Extrasystoles
IV	Purkinje (4 of 11)	150	27	90	2:1 Block, Oscillatory Conduction
V	Purkinje (4 of 8)	150	37	240	2:1 Block Oscillatory Conduction

Numbers below tissue indicate frequency of recorded arrhythmias in each series. In I and III arrhythmias were observed but not routinely recorded in every experiment.
* see Ornhagen and Hogan (1977).

Conduction arrhythmias ranging in severity from coupled extrasystoles to complete conduction block have been demonstrated in tissue studies (Table 3) at levels of pressure that are now attainable by man. These disturbances are exacerbated by low temperature and rapid heart rate, conditions that might be expected to obtain when the diver is at work during prolonged exposure in the hyperbaric environment. It is encouraging to find, in this regard, that the untoward effects of pressure on cardiac conduction are ameliorated by narcotic gases of the kind that are presently included in the breathing mixtures used in very deep dives.

III. HYDROSTATIC PRESSURE EFFECTS ON CARDIAC CONTRACTION

A unique feature of cardiac cells is their ability to adjust their contractile force commensurate with the demand for cardiac output. The effect of pressure on the contractile state of the heart was first described by Cattell and Edwards (1929), Edwards and Cattell (1930), and later by Brown (1957). Our recent studies are in agreement with this earlier work in demonstrating that cardiac contractility is one of the most pressure-sensitive processes of all biological phenomena. We have calculated an apparent activation volume for the twitch contraction of the cat papillary muscle of about $-200\ cm^3.mol^{-1}$. As shown in Table 2, this value is considerably larger than those for electrical phenomena and more importantly is a negative value indicating that elevated pressure enhances contractile force. Fig.8 demonstrates the important fact that the relationship between pressure and twitch tension is itself dependent upon the beating frequency of the heart. In general, the elevated twitch tension seen at faster beating frequencies is less sensitive to a change in pressure. In the extreme of very rapid frequency, very high pressure may in fact depress contraction force.

The contractile state of cardiac muscle falls to zero at the end of each contraction, returning to its steady-state value during the subsequent diastolic interval. The restitution of contractility during diastole has provided us with a means for assessing the mechanism underlying the influence of pressure on cardiac contractility. The typical effect of pressure on the restitution curve is shown in Fig.9. Relevant changes in the curve at 200 atmospheres are 1) a marked increase in the upper asymptote, 2) a decrease in the time constant of the ascending limb of the curve, and 3) a decrease in the time constant for the falling limb of the curve.

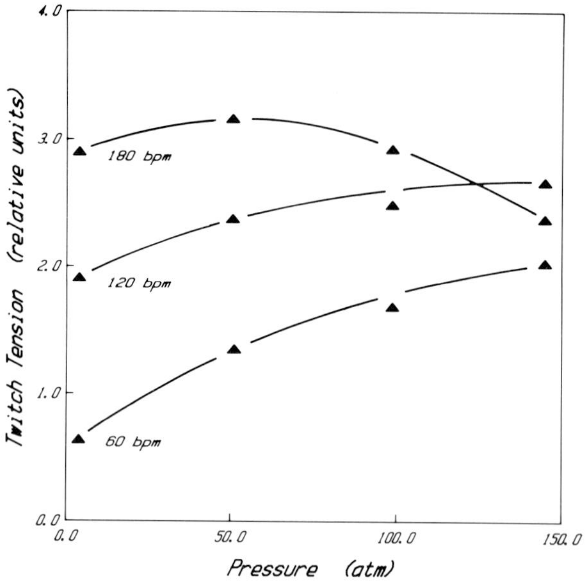

Fig.8. *Effect of elevated hydrostatic pressure and stimulation frequency on twitch tension in cat papillary muscle.*

These findings compare favorably with the response of a four-compartment computer simulation of calcium turnover in heart cells, in which the kinetics for calcium reuptake by the internal storage sites and calcium extrusion via the sodium-calcium exchanger were markedly slowed. While this may be fortuitous, it is intriguing that both of these transport processes depend upon complex protein carrier molecules of the kind that well might be expected to respond to changes in pressure. The net effect of these changes is an increase in the intracellular calcium available for activation of the contractile filaments. This putative mechanism would provide a distinct advantage in maintaining cardiac performance in response to the stress of the hyperbaric environment.

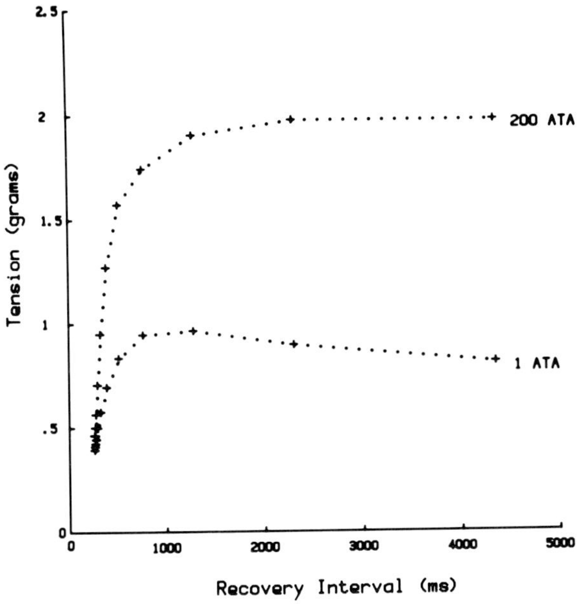

Fig.9. *Effect of elevated hydrostatic pressure on twitch tension restitution kinetics in rabbit papillary muscle.*

Acknowledgements. The author extends his gratitude to Drs. Thomas Doubt and Hans Ornhagen, who collaborated in this work and to Jacqueline Dahl for her invaluable technical support. This work was sponsored by USPHS grants PO1 HL 28542-02.

REFERENCES

Brauer, R.W., Hogan, P.M., Hugon, M., MacDonald, A.G., Miller, K.W. (1982) Patterns of interaction of effects of the light metabolically inert gases with those of hydrostatic pressure as such - a review. Undersea Biomed. Res. 9(4), 353-396.

Brown, D.E.S.(1957) Temperature-pressure relation in muscular contraction. In : Johnson, F.H. (ed), Influence of temperature on biological systems. American Physiological Society, Washington, D.C.

Cattell, M., Edwards, D.J.(1929) The influence of hydrostatic pressure on the contraction of cardiac muscle in relation to temperature. Am. J. Physiol. 93, 97-104.

Conti, F., Fioravanti, R., Segal, J.R., Stuhmer, W.(1982) Pressure dependence of the sodium currents of squid giant axon. J. Memb. Biol. 69, 23-34.

Doubt, T.J., Hogan, P.M.(1978) Effects of hydrostatic pressure on conduction and excitability in rabbit atria. J. Appl. Physiol. 45(1), 24-32.

Doubt, T.J., Hogan, P.M. (1979). Action potential correlates of pressure-induced changes in cardiac conduction. J. Appl. Physiol. 47(6), 1169-1175.

Doubt, T.J., Hogan, P.M.(1981) The arrhythmogenic potency of hydrostatic pressure on cardiac conduction. In : Bachrach, A.J., Matzen, M.M.(eds) Underwater Physiology VII. Proceedings of the Seventh Symposium on Underwater Physiology. Undersea Medical Society, Inc., Bethesda MD. pp. 235-240.

Doubt, T.J., Hogan P.M.(1982) Combined effect of beating rate and hydrostatic pressure on excitation in cardiac muscle. Undersea Biomed. Res. 9(3), 241-253.

Edwards, D.J., Cattell, M.(1930) The action of compression on the contraction of heart muscle. Am. J. Physiol. 93, 90-96.

Goldinger, J.M., Kang B.S., Choo, Y.E., Paganelli, C.V., Hong, S.K. (1980) Effect of hydrostatic pressure on ion transport and metabolism in human erythrocytes. J. Appl. Physiol. 49(2), 224-231.

Johnson, S.M., Miller, K.W. (1975) The effect of pressure and volume of activation on the monovalent cation and glucose permeabilities of liposomes of varying composition. Biochem. Biophys. Acta, 375, 286-291.

Ornhagen, H.C.(1979) Influence of nitrous oxide, nitrogen, neon, and helium on the beating frequency of the mouse sinus node at high pressure. Undersea Biomed. Res. 6(1), 27-39.

Ornhagen, H.C., Hogan, P.M.(1977) Hydrostatic pressure and mammalian cardiac pacemaker function. Undersea Biomed.Res.4(4), 347-358.

Wann, K.T., MacDonald, A.G.(1980) The effects of pressure on excitable cells. Comp. Biochem. Physiol. 66A, 1-12.

FROM SIMPLE TO COMPLEX : PRESSURE AND INERT GAS MODULATION OF NERVE CELL MEMBRANE PROPERTIES

J.J. KENDIG.

I. INTRODUCTION

When subjected to high hydrostatic or helium pressure of approximately 2 - 10 MPA (20 - 100 atmospheres), vertebrates display symptoms of heightened neural excitation manifested as the High Pressure Neurological (or Nervous) Syndrome (HPNS) (Brauer, 1984). The syndrome includes, among other manifestations, tremors and convulsions. In anesthetized animals, an additional stimulatory effect of pressure is manifested as an increase in the amount of anesthetic required to prevent mobility (Miller, 1977). There are actually four different CNS responses to inert gases and pressure : (1) anesthesia (and inert gas narcosis); (2) the convulsions and tremors of the HPNS (3) the anti-anesthetic effects of pressure, and (4) the anti-HPNS effects of anesthetics. The cellular and molecular sites of HPNS and of pressure reversal of anesthesia are considered to be different because of dissimilarities in pharmacology and in pressure-anesthetic dose/response curves (Halsey, 1982), and because of the limited window of anesthetic and pressure exposure in which animals are both awake and seizure-free (Miller 1977). For neither phenomenon is the cellular basis known.

A major unresolved question is whether anesthesia and pressure antagonism to anesthesia involve the same site, as is implied in the poorly chosen phrase "pressure reversal of anesthesia", or whether this is an indirect antagonism. The question is an important one. If the antagonism between pressure and anesthetics consists of directly opposing effects at the same molecular site(s), then pressure can be used as a tool to probe the nature of anesthesia. If the antagonism is indirect, it is still interesting but rather less informative. A similar question can be asked for responses (2) and (4), i.e., whether the HPNS-ameliorative effects of anesthetics constitute a direct antagonism to pressure.

II. REVIEW OF NEURONAL RESPONSES TO PRESSURE

The effects of pressure can be assumed to be ubiquitous, affecting all excitable cells. The most sensitive areas, those responsible for HPNS symptoms and pressure-anesthetic antagonism, have not been clearly identified. Spinal animals can display some of the symptoms of HPNS, and decerebrate animals exhibit motor convulsions on exposure to pressures lower than those required to initiate seizures in intact animals (Angel *et al.*, 1984). The increased sensitivity to pressure of decerebrate animals has been cited in support of the suggestion that the cerebral cortex exerts an inhibitory effect on pressure-induced convulsions (Angel *et al.*, 1984).

Most of the information on HPNS has been derived from the pharmacology of HPNS antagonists. Because a weakening of inhibitory circuits is strongly implicated in other models of seizure disorders, depression of inhibition is an obvious candidate for an explanation of HPNS seizures. Specific agonists and antagonists of GABA-ergic and glycinergic inhibitory synaptic transmission have been tested for efficacy in altering the pressure threshold for HPNS seizures. Most GABA agonists and partial agonists do raise the seizure threshold in HPNS, whereas picrotoxin lowers the threshold (Bowser-Riley, 1984). Glycinergic inhibition is implicated by a recent study which shows that strychnine and hyperbaric pressure are linearly additive in seizure production (Smith and Bowser-Riley, 1984), presumably at some subcortical site. A few studies have been done on cortical evoked potentials at hyperbaric pressure. In intact animals, both somatosensory and visual evoked potentials have been reported to increase on compression (Angel *et al.*, 1980; Kaufmann *et al.*, 1981), as they are reported to do in other convulsive states. The underlying mechanism for the increase is not known.

Although a fair number of studies have now been carried out in nerve cells exposed to hyperbaric pressure, surprisingly little support has emerged from them for any particular cellular basis for either HPNS or pressure reversal of anesthesia. Synaptic transmission was one of the first processes to be examined at hyperbaric pressure. Intracellular studies have been carried out in squid giant synapse (Henderson *et al.*, 1977), at the vertebrate neuromuscular junction (Ashford *et al.*, 1982), in molluscan ganglion cells (Parmentier *et al.*, 1981), and at the crayfish neuromuscular junction (Campenot, 1975). The common finding in these disparate systems supports an initial result, obtained with extracellular recording in vertebrate sympathetic ganglia (Kendig *et al.*, 1975), showing that pressure depresses excitatory synaptic transmission. At all the excitatory synapses so far examined directly, pressure depresses transmission. In two synaptic preparations studied in more detail, the vertebrate neuromuscular junction and molluscan cell bodies, pressure-induced decrease in MEPP frequency and absence of pressure effects on the response to iontophoretically applied transmitter point to a decrease in transmitter release as the cause of the depressant effect of pressure. In the absence of any indication that pressure can enhance rather than depress any type of excitatory transmission, it is difficulty to try to fit this finding in with the "excitatory" effect of pressure as convulsant and anti-anesthetic.

A major gap in our knowledge is the lack of studies on inhibitory synapses at pressure. A selective depressant effect of pressure on inhibitory transmission has been proposed on the indirect pharmacologic evidence outlined above, but such an effect has yet to be demonstrated directly at any inhibitory synapse. To account for pressure-induced hyperexcitability, a selective depression of inhibitory as opposed to excitatory synapses would have to be postulated; it is not clear what the basis of such a selectivity might be.

The findings cited above provide no support for the possibility that pressure augments neural excitation by enhancing transmission at excitatory synapses. Because pressure can reverse partial conduction block in axons (Kendig *et al.* , 1975), recent studies in our laboratory have focussed on the effects of pressure on the conducted action potential and its underlying voltage-dependent sodium and potassium channels. We have exposed axons of several different types to hyperbaric pressure. The results from our studies and those of other laboratories show that pressure affects fundamental channel properties qualitatively in a similar fashion from axon to axon. However, the outcome in terms of ability to conduct an action potential varies considerably.

III. METHODS

Axons of two different structural types were chosen for comparison, namely vertebrate myelinated peripheral nerve and an invertebrate branching motor neuron. The latter, although unmyelinated, has a very elaborate sheath. It also has the interesting property of lability in conduction; the axon fails to pass impulses at its branch point after several seconds of firing at physiological frequencies of

30-50 Hz (Spira *et al.*, 1976). Large vertebrate myelinated axons have a very much higher safety factor, being capable of conducting impulses at up to 500 Hz for prolonged intervals.

Axons from frog (*Xenopus laevis*) sciatic nerve were prepared for vaseline gap voltage clamping by our standard procedure (Kendig *et al.*, 1979). In most cases sodium and potassium currents were pharmacologically separated by blocking the axon with tetraethylammonium chloride (TEA) or tetrodotoxin (TTX) respectively. The chamber containing the node of Ranvier was placed on a Peltier thermoelectric device within a small pressure chamber. Temperature was maintained at approximately 10° ± 0.25°C while measurements were made; transient excursions during compression and decompression steps were limited to 2°C. The chamber was pressurized by admitting helium from a commercial cylinder at 3500 psi; compression was carried out in steps of 3.5 MPa.

The bifurcating common excitor motor neuron to the dorsal abdominal extensor muscles of the lobsters *Homarus americanus* or *Panulirus interruptus* was isolated according to published methods (Grossman *et al.*, 1979). The axon (Ax) and its two branches (M, medial, and L, lateral) were prepared for extracellular stimulation and recording in the small pressure chamber, using modified suction electrodes. A limited number of voltage clamp experiments on the main axon segment were also carried out in this chamber, using the same techniques as for the node of Ranvier with some modification. Compression and temperature control were as described above. A larger chamber, accommodating a remote-controlled micromanipulator, was used to permit simultaneous extracellular microelectrode recording from Ax. The preparation was usually maintained at room temperature (20-22°C), with some experiments at lower temperatures.

IV. RESULTS

In the node of Ranvier, there was some irreversible loss of sodium current on compression. The most striking finding was a marked slowing of the rate of rise and decay of the current (Fig.1). There was no consistent change in the current/voltage relationship. There was, however, a marked consistent shift in the voltage dependence of inactivation in the depolarizing direction, which corresponded to an increase in sodium current amplitude at holding potentials where the control inactivation constant (Hodgkin-Huxley h) was less than 1 (Fig.2).

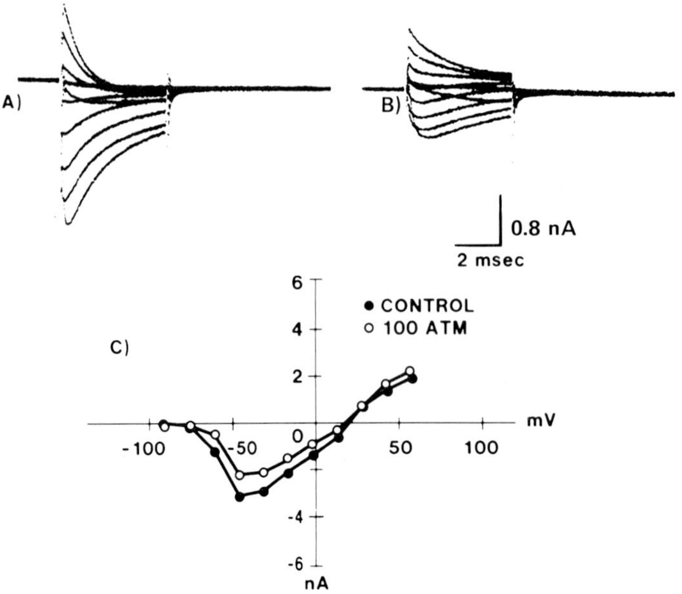

Fig.1. Sodium currents in the node of Ranvier at 1(a) and 100 atmospheres (b). Response to a series of depolarizing voltage steps of 7 msec duration at 15 mV increments from a holding potential of 83 mV. Potassium channels blocked with TEA. Currents are slowed at all potentials. Leakage current, electronically subtracted, did not change on compression. The effect of pressure on current time course was partially reversed on decompression. Current/voltage relation shape for a different node (c) was not altered by compression, although currents were smaller after compression. From Kendig (1984a), with permission.

The shift in inactivation appeared to be related to a selective effect of pressure on the rate of development of inactivation. When τ_h is plotted as a function of membrane voltage, there appear to be large pressure effects in the range in which β_h is the predominant rate constant, but only small effects in the more hyperpolarized range (Fig.3). These effects of pressure were measurable as soon as temperature had returned to control following a compression step and were stable over time up to thirty minutes at a given pressure. Potassium currents also exhibited a marked decrease in rate of rise.

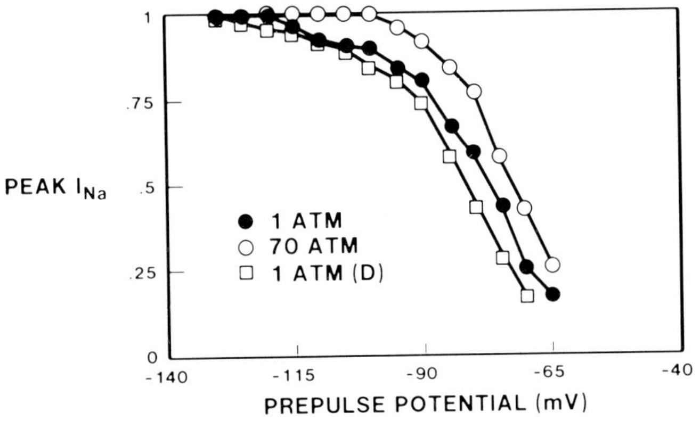

Fig. 2. Shift in the voltage dependence of steady-state inactivation in the node of Ranvier at 70 atmospheres. Vertical scale, peak sodium current referred to the maximum sodium current elicited at hyperpolarized potentials (h_∞); horizontal scale, amplitude of a 50 msec voltage step preceding a depolarizing test step to - 23 mV. The shift in h_∞ observed at 70 atmospheres is reversed on decompression to 1 atmosphere. From Kendig (1984a) with permission.

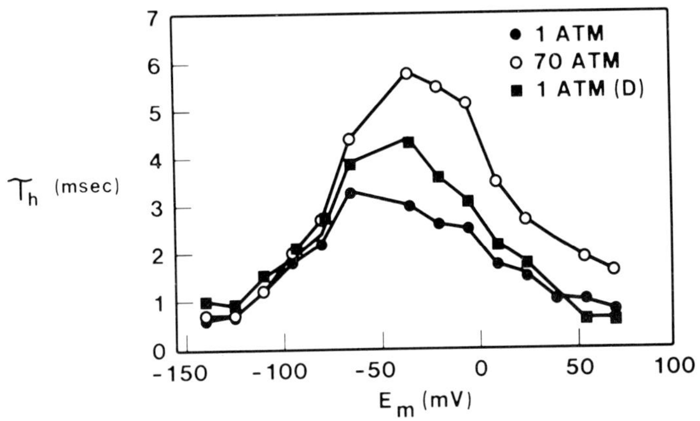

Fig. 3. τ_h versus membrane voltage in the node of Ranvier at 1 and 70 atmospheres, and on decompression to 1 atmosphere. τ_h was measured by a 3-pulse protocol at potentials negative to - 50mV, and as sodium current decay at more positive potentials. The increase in τ_h following compression is voltage-dependent, being more pronounded at the more depolarized potentials. Recovery of τ_h on decompression was incomplete. From Kendig (1984a) with permission.

By contrast, the effect of pressure on the crustacean bifurcating axon were complex and time-dependent. Initially, during the first fifteen minutes following a compression step, there was an increase in conducting ability as measured by increases in extracellularly recorded action potential amplitude, conduction velocity, and ability to conduct at high frequencies (Fig.4).

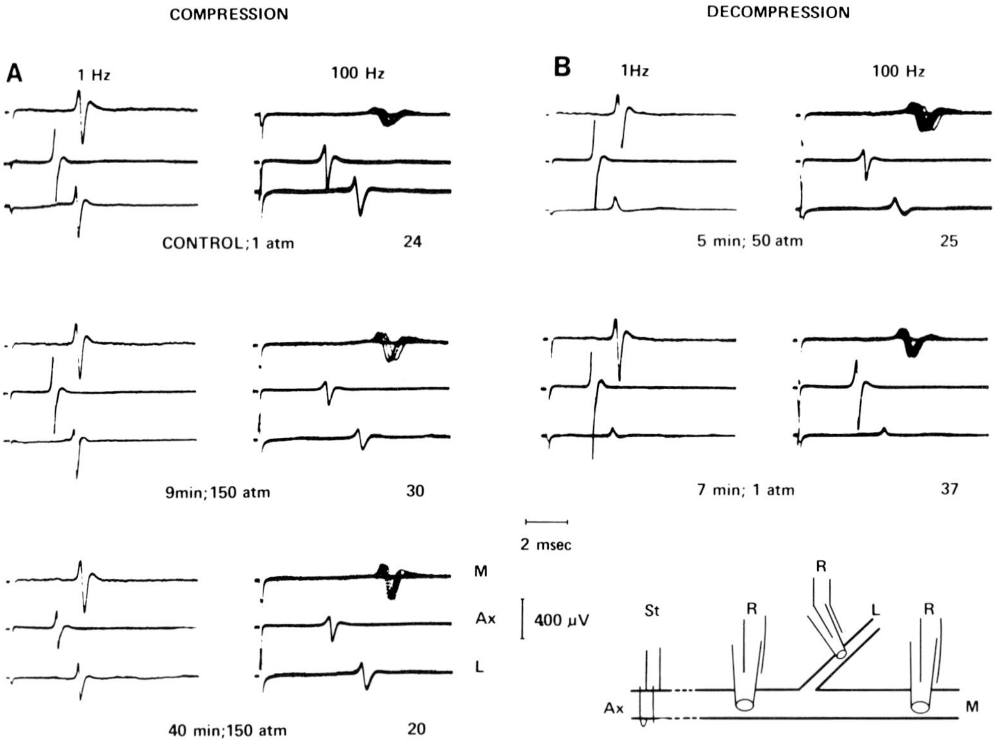

Fig.4. *Initial and delayed effects of pressure changes on the extracellularly recorded action potential of a bifurcating lobster motor neuron. Inset diagram shows the recording arrangement: St, stimulating electrodes, R, recording electrodes, Ax, main axon; M,medial, and L, lateral branch. Left traces, responses in M, Ax and L at 1 Hz; right traces at 100 Hz. Figures beneath right traces are seconds required to block impulses in M at 100 Hz at each pressure. A, compression produces an initial increase in time required to block at 100 Hz, followed by a delayed decrease when the preparation is held at the same pressure. B, similar initial increases in conducting ability were observed following decompression steps.*

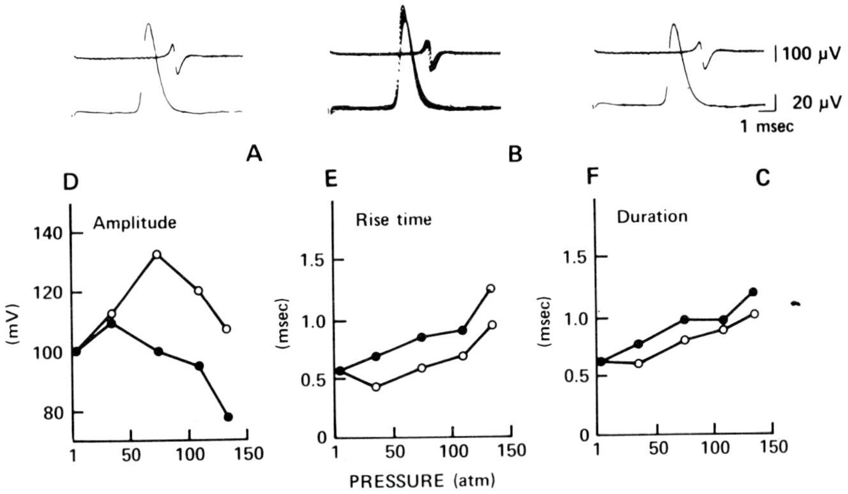

Fig.5. Effects of pressure on the intracellularly recorded action potential of the common excitor motor neuron of the dorsal abdominal extensor muscles in the abdomen of Panulirus interruptus. A-C, upper traces, extracellularly recorded responses in the M(medial) branch; lower traces, intracellularly recorded responses in Ax (main axon). A, stable response at 35 atm; B, superimposed responses elicited at 1 Hz during compression to 70 atm; the response increases in amplitude, rise time is faster, and conduction velocity increases. C, after 20 minutes at 70 atm, these measures have relaxed to values below those at 35 atm. D-F, maximum initial ○ and delayed ● (20-27 min) responses of the same axon on repeated compression, showing initial changes in direction opposite to the delayed effects. From Grossman and Kendig (1984), with permission.

The intracellularly recorded action potential also clearly showed an increase in amplitude and rate of rise (Fig.5). Voltage clamp recordings showed a marked increase in the amplitude of the inward sodium current; increases in time constants of both inward and outward currents were present, but small compared to those observed in vertebrate myelinated axons (Fig.6). There was suggestive evidence that the increase in sodium current amplitude was related to a depolarizing inactivation shift similar to that observed in the node of Ranvier, but the number of voltage clamp experiments was too small to be certain. However, the increases in these measures of excitability proved to be transient. After twenty minutes at a given pressure, the initial increases in conducting ability and sodium channel function were succeeded by declines to below control values in all measures (Fig. 4-6). In addition, at pressures above 3.5 MPa, there was a slowly developing depolarization which reached a maximum of 10-15 mV.

Depolarization was observed in all axons except those pretreated with ouabain to inhibit active sodium transport.

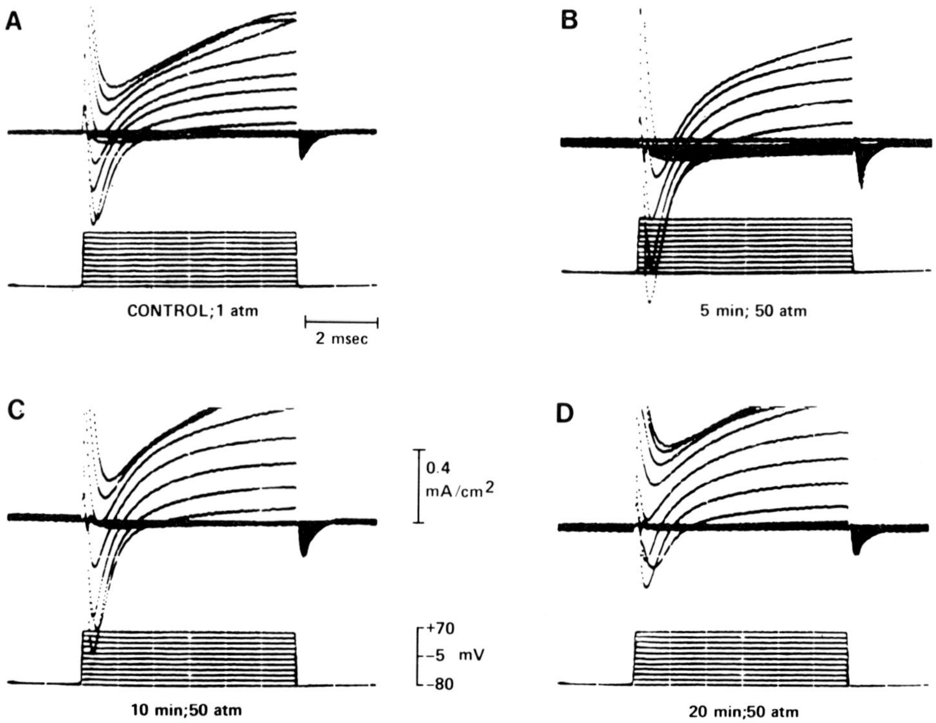

Fig.6. Initial and delayed effects of pressure on currents in the voltage-clamped Ax segment of the lobster motor neuron. Upper traces, currents; lower traces, voltage steps. Time at 50 atm noted beneath each set of currents. From Grossman and Kendig (1984), with permission.

V. DISCUSSION

In myelinated vertebrate axons, as well as in squid giant axons and crustacean giant axons, part of the conduction block induced by all general anesthetics (except short chain alcohols) is due to a hyperpolarizing shift in the voltage dependence of inactivation, which thus reduces the availability of sodium channels at the normal resting potential (Bean *et al.*, 1981; Haydon and Urban, 1983; Kendig, 1981; Kendig *et al.*, 1979). We have shown that this property also applies to the inert gas nitrogen at hyperbaric pressures, supporting the often unspoken assumption that organic volatile anesthetics and inert gases interact with nerve membranes in fundamentally similar fashion (Kendig, 1984a,b). The opposing shifts in inactivation voltage dependence observed at hyperbaric pressure and in the presence of anesthetics respectively would support exactly the sort of limited pressure relief of anesthetic conduction block we have seen in intact nerve bundles : a partial reversal; since in addition to this voltage dependent component of block, there is a basal block of sodium channels which can only be relieved to a limited extent by this mechanism.

Our strategy in choosing to examine for comparison the branching crustacean axon in which conduction is labile was based on the sensitivity of this axon to anesthetic agents. At close to clinical concentrations, anesthetics such as diethyl ether and halothane can dramatically shift downwards the frequency response of the axon, thus reducing its ability to pass impulse trains (Grossman and Kendig, 1982). It was our hypothesis that if this axon supported pressure antagonism to the anesthetic effect, then it might serve as a model for a central axon or impulse-conducting cell segment which could display the requisite sensitivity to be considered a plausible site of anesthesia. As shown above, increases in excitability on compression do occur in this axon, and would give rise to alleviation of the effects of anesthetics. However, their transient nature and the succeeding marked long-term depression, observed by three different experimental techniques, demonstrates that pressure reversal of conduction block would not obtain as a steady-state phenomenon in this axon. An additional component of depression, not observed in other axons, appears to be due to depolarization following inhibition of active ion transport.

A third type of axon, the well-studied squid giant axon, has also been examined at hyperbaric pressure in a detailed fashion. We have heard evidence today that the individual currents in this axon do not support pressure reversal of conduction block. This is confirmed by other studies which show no shift in sodium channel steady-state inactivation at even very high pressures, although qualitatively there is a selective slowing of inactivation development, but very much smaller than that observed in myelinated nerve (Conti *et al.*, 1982).

Thus, there are examples of three different axons , one of which supports pressure reversal of anesthetic conduction block by properties inherent in the sodium channel (myelinated nerve); a second which would support pressure reversal of conduction block also based on sodium channel behavior, but of a transient nature, succeeded by a depression whose nature is not clear, but which involves resting potential changes in part (crustacean nerve); and a third, squid giant axon, which has been reported to support pressure reversal of block of the conducted action potential which, if substantiated, is not due to sodium channel properties alone. In addition, a small population of crustacean peripheral axons appears to respond to compression by generating repetitive impulse activity (Kendig *et al.*, 1978).

Thus, although qualitatively slowing of ion channel rate constants is a fundamental effect of pressure on the voltage-dependent sodium and potassium channels, quantitatively the magnitude of the effect varies considerably. When translated to the integrated ability of the membrane to conduct an action potential, pressure effects on axonal membrane are strongly determined by the axon environment in which the evolutionarily conservative sodium channel resides. The implication of these findings is to restrict the extent to which results from easily accessible peripheral axons can be generalized to axons or other impulse-conducting membrane in the central nervous system. If there is a role for the voltage-dependent sodium channel in the central symptoms of exposure to hyperbaric pressure, it will not be established until central neurons are examined.

REFERENCES

Angel, A., Gratton, D., Halsey, M.J., Wardley-Smith, B.(1980) Pressure reversal of the effect of urethane on the evoked somatosensory cortical response in the rat. Br. J. Pharmacol. 70, 241-247.

Angel, A., Halsey, M.J., Little, H., Meldrum, B.S., Ross, J.A.S., Rostain, J.C., Wardley-Smith, B.(1984) Specific effects of drugs at pressure : animal investigations. Phil. Trans. Roy. Soc. Lond,B304, 85-94.

Ashford, M.L.J., MacDonald, A.G., Wann, K.T.(1982) The effect of hydrostatic pressure on the spontaneous release of transmitter at the frog neuromuscular junction. J. Physiol. 333, 531-543.

Bean, B.P., Shrager, P., Goldstein, D.A.(1981) Modification of sodium and potassium channel gating kinetics by ether and by halothane. J. Gen. Physiol. 77, 233-253.

Bowser-Riley, F.(1984) Mechanistic studies on the high pressure nervous syndrome. Phil. Trans. Roy. Soc.Lond. B, 304, 31-41.

Brauer, R.W.(1984) Hydrostatic pressure effects on the central nervous system : perspectives and outlook. Phil. Trans. Roy. Soc. Lond. B 304, 17-30.

Campenot, R.B. (1975) The effects of high hydrostatic pressure on transmission at the crustacean neuromuscular junction. Comp. Biochem. Physiol. B52, 133-140.

Conti, F., Fioravanti, R., Segal, J.R., Stühmer, W. (1982) Pressure dependence of the sodium currents of squid giant axon. J. Memb. Biol. 69, 23-34.

Grossman, Y., Kendig, J.J. (1982) General anesthetic block of a bifurcating axon. Brain Res. 245, 148-153.

Grossman, Y., Kendig, J.J.(1984) Pressure and temperature : time-dependent modulation of membrane properties in a bifurcating axon. J. Neurophysiol. In press.

Grossman, Y., Parnas, I., Spira, M.E.(1979) Ionic mechanisms involved in differential conduction of action potentials at high frequency in a branching axon. J. Physiol. 295, 307-322.

Haydon, D.A., Urban, B.W.(1983) The action of hydrocarbons and carbon tetrachloride on the sodium currents of the squid giant axon. J. Physiol. 338, 435-450.

Halsey, M.J.(1982) Effects of high pressure on the central nervous system. Physiol. Rev. 62, 1341-1377.

Henderson, J.V., Lowenhaupt, M.T., Gilbert, D.L.(1977) Helium pressure alteration of function in squid giant synapse. Undersea Biomed. Res. 4, 19-26.

Kendig, J.J.(1981) Barbiturates : active form and site of action at node of Ranvier sodium channels. J. Pharmacol. Exp. Ther. 218, 175-181.

Kendig, J.J. (1984a) Ionic currents in vertebrate myelinated nerve at hyperbaric pressure. Am. J. Physiol. 246, C84-C90.

Kendig, J.J. (1984b) Nitrogen narcosis and pressure reversal of anesthetic effects in node of Ranvier. Am. J. Physiol. 246, C91-C95.

Kendig, J.J., Courtnez, K.R., Cohen, E.N.(1979) Anesthetics : molecular correlates of voltage and frequency dependent sodium channel block in nerve. J. Pharmacol. Exp. Ther. 210, 446-452.

Kendig, J.J., Schneider , T.M., Cohen, E.N.(1978) Pressure, temperature and repetitive impulse generation in crustacean axons. J. Appl. Physiol. 45, 742-746.

Kendig, J.J., Trudell, J.R., Cohen, E.N.(1975) Effects of pressure and anesthetics on conduction and synaptic transmission. J. Pharm. Exp. Ther. 195 , 216-224.

Miller, K.W. (1977) The opposing physiological effects of high pressure and inert gases. Fed. Proc. 36, 1663-1667.

Parmentier, J.L., Shrivastav, B.B., Bennett, P.B.(1981) Hydrostatic pressure reduces synaptic efficiency by inhibiting transmitter release. Undersea Biomed. Res. 8, 175-183.

Smith, E.B., Bowser-Riley, F. Pressure reversal and the pharmacology of the high pressure neurological syndrome. Molecular Mechanisms of Anesthesia, Plenum, NY. In press.

Spira, M.E., Yarom, Y., Parnas, I. (1976) Modulation of spike frequency by regions of special axonal geometry and by synaptic inputs. J. Neurophysiol. 36, 882-899.

THE ROLE OF PERIPHERAL NEUROMUSCULAR ELEMENTS IN THE TREMOR-RELATED SYMPTOMS OF THE HIGH PRESSURE NERVOUS SYNDROME

D.J. HARRIS, J.L. PARMENTIER.

I. INTRODUCTION

All animal species show characteristic increases in their nervous system excitability when subjected to hydrostatic compression. This may be expressed initially as a heightened level of normal behavioral activity; then, as compression increases, more overt motor signs may be observed including tremors, myoclonus and seizures. In man the emotional distress brought about by "vestibular" EEG and sleep disturbances is compounded by symptoms of uncoordination, tremors, muscle jerks and fatigue which limit performance. In seeking to understand these symptoms, the hyperbaric physiologist has not been helped by the controversy in the literature on the origin(s) of normal (physiological action) tremor, let alone the various forms of pathological tremor.

In 1980, at the 28th International Congress of Physiological Sciences in Budapest, the concluding remarks in the session on the effects of pressure on the mammalian nervous system included the following statement : "the High Pressure Nervous Syndrome (HPNS) was not a single entity but a compound one, resulting from the intertwining of several concurrent processes which lead up to different endpoints ... characterising the HPNS" (Hugon and Brauer, 1981). It was further suggested that "HPNS-like phenomena are manifestations of nerve-network characteristics and hence not predictable in terms of the isolated neurone". Attention was drawn to the gap between work at the cellular level or below and studies of intact organisms.

In recent years laboratories in France, the UK and the USA have been working to fill this gap by studying the effects of pressure on spinal reflex-loop excitability, particularly that of the knee-jerk reflex, the Achilles tendon reflex and the Hoffmann reflex. The anatomy and physiology of these monosynaptic reflex loops make then convenient neuronal subsystems for study *in vivo* since responses can be easily and painlessly elicited, and measured without using invasive techniques. Systematic reflex studies have shown that there are definite alterations in the excitability of the stretch reflex pathway and in the way muscles contract *in vivo* at pressure. It now appears that changes in these peripheral components of the neuromuscular system may be crucial to the propagation of high pressure tremor, and confirm their importance in the generation of normal (physiological) tremor.

II. HYPERBARIC HYPERREFLEXIA

A. Stretch reflexes

Considered simply, these reflexes involve only two neurones, an afferent and an efferent. The afferent neurone separates a sensory input transducer (the muscle spindle) from a central gain control (synapse plus interneurones). The efferent neurone connects the central gain control with a motor output device (the muscle fibre). In the familiar ankle jerk, a brisk stretch of the Achilles tendon excites the muscle spindles in the belly of the calf muscles which in turn initiates a discrete afferent volley. In the cord, this volley impinges on the alpha motoneurones, the efferent volley then resulting in a single twitch of the soleus and/or gastrocnemius muscles. The responsiveness of such myotatic or "T" reflexes depends upon the tone of the muscle spindles which are under supraspinal fusimotor (gamma efferent) control, and upon the excitability of the alpha motoneurones which, in turn, are influenced by both spinal feedback and supraspinal control pathways.

Using specially designed reflex chairs to carefully control and standardise the subjects'posture (Desmedt, 1973; Hugon, 1973), electromyographic (EMG) studies of the T-reflexes in both quadriceps and soleus/gastrocnemius muscles of many deep divers have confirmed that stretch reflexes are facilitated by 20-80% during compression to 250 msw or more (Fig.1) (Bonnet *et al.*, 1973; Roll *et al.*, 1978; Harris, 1979a,b; Hugon *et al.*, 1983; Harris and Bennett, 1984a; Parmentier *et al.*, 1984). Interestingly, the effect of Jendrassik's manoeuvre (an arm-pull or fist-clench during reflex stimulation which will readily facilitate the stretch reflex by 10-30% at the surface) was abolished at pressure (Roll *et al.*, 1978; Harris, 1979a, b). Hyperbaric hyperreflexia is invariably associated with increased tremor and decreased motor performance. But whereas hyperreflexia often returns at the end of a dive and persists into the post-dive period, tremor does not (Roll *et al.*, 1978; Harris, 1979b; Parmentier *et al.*, 1984).

The EMG records from deep divers also reveal abnormal EMG potentials. Low frequency, large amplitude potentials may be associated with muscle fasciculations; they do not result in any coordinated muscle activity. Much smaller potentials often occur at a reproducible delay after a reflex discharge. These polyphasic discharges were initially felt to polysynaptic reflexes, and they were compared with motor "startle" reactions (Roll *et al.*, 1978; Harris, 1979b). EMG potentials of this kind which have been observed in recent dives appear to coincide exactly with the rebound of the recovery cycle (see below) and are probably secondary stretch reflexes resulting from monosynaptic afferent re-excitation.

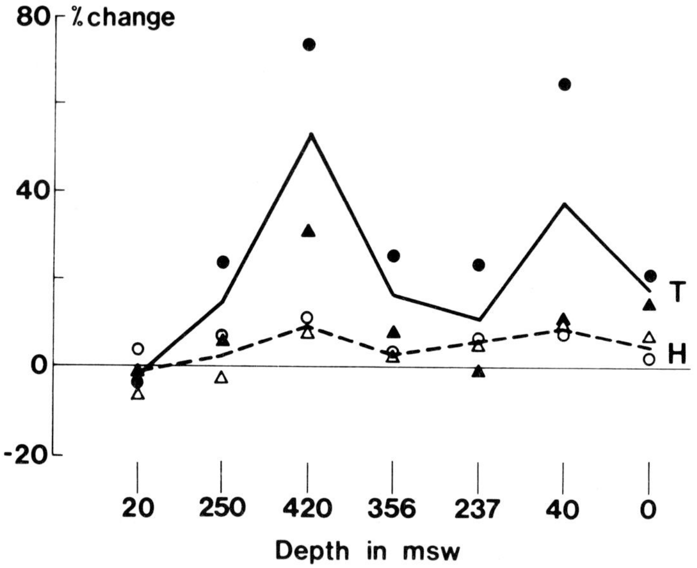

Fig.1. Mean change in the recruitment ratios of soleus T-reflex (solid line, filled symbols) and H-reflex (dashed line, open symbols) for 2 subjects (circles, triangles) during a 420 msw heliox saturation dive at AMTE(PL) (Dive 8,1978). Vertical axis shows percentage difference from pre-dive controls. After Harris (1979b) modified.

B. The Hoffmann Reflex

On the evidence so far, hyperreflexia in divers could be due to either an increase in muscle spindle sensitivity, or a change in alpha moto-neurone excitability, or both. To try and answer this question, the Hoffmann reflex (H-reflex) has also been studied. The H-reflex is a response similar in electromyographic appearance to the T-reflex but which is obtained by percutaneous electrical stimulation of the muscle spindle afferents in the popliteal nerve. The neuromuscular circuitry of the H-reflex is identical to that of the T-reflex except that the muscle stretch receptors are bypassed. H-reflex excitability therefore depends solely on the excitability of the alpha motoneurone pool, other factors being equal (Hugon, 1973; Delwaide, 1976). In a comparative study of T- and H-reflexes in the same muscle, a change in responsiveness in both reflexes can only be due to central effects on alpha motoneurone excitability; an increase in the T-reflex alone could only be due to a stretch-receptor effect.

When H-reflex excitability was measured at pressure in heliox it was found that it was facilitated at pressure, but only slightly when compared with the facilitation of the T-reflex (Roll *et al.*, 1978; Harris, 1979b). Fig.1 shows the change in H-reflex and T-reflex excitability of the soleus in two subjects during a 420 msw dive at AMTE(PL) (now ARE(PL)), Gosport, England. This differential facilitation of T- and H-reflexes has been noted in man on several other deep dives (Roll *et al.*,1978; Parmentier *et al.*, 1984), and in the Baboon at 1000 msw (Bonnet *et al.*, 1973) and the monkey (Hugon *et al.*, 1981) as well.

If the underlying rationale is correct, the absence of hyperreflexia in the H-reflex at pressures which facilitate the T-reflex strongly suggests that the main cause of hyperbaric hyperreflexia lies in an increased sensitivity of the muscle spindles, and that only a small part is due to central synaptic facilitation of the alpha motoneurones. Unfortunately, there are some grounds for doubting the equivalence (and thus the comparability) of the H-and T-reflex volleys. Because the stimulus is so brief, the H-reflex volley appears to be far more synchronised spatially and temporally than the T-reflex volley, making it potentially resistant to subtle or graded synaptic inputs (Delwaide and Delbecq, 1973; Burke *et al.*, 1983). Furthermore, there is recent evidence that popliteal stimulation adequate for the Ia afferent nerves from soleus is also supraliminal for afferents from other muscles in the sale limb and from cutaneous mechanoreceptors, as well as for Ib afferents. The afferent population excited in this way is so different from that of the tap-elicited T-reflex volley that direct comparison of T- and H-reflexes for elucidating stretch receptor function is probably invalid (Burke *et al.*, 1983).

In summary, compression elicits large increases in stretch reflex responsiveness, relatively small increases in H-reflex responsiveness, and abolishes the effectiveness of Jendrassik's manoeuvre. However, it is not possible to make definitive statements on the mechanism of hyperreflexia based on these facts because of uncertainty about the normal mechanism of the Jendrassik effect and the probable differences between the H- and the T-reflex volleys. The picture becomes clearer when other, more complex, measurements of H-reflex function and simple observations on muscle twitch contractions are considered.

III. H-REFLEX RECOVERY CYCLE OF EXCITABILITY

The time course of central excitability changes which follow a reflex discharge can be determined by comparing the amplitude of the H-reflex responses to a pair of popliteal stimuli of equal strength (S1, S2), each of which would normally elicit a single H-reflex of 50% maximum amplitude. Stimuli are delivered every 10 seconds with an inter-stimulus delay ranging from 10-5000 ms. Changes in excitability of the reflex pathway following the first (H1) discharge are reflected in changes in the amplitude of the second (H2) response (Paillard, 1955; Hugon, 1973; Delwaide *et al.*, 1976). Paillard noted five phases of the recovery cycle of which three have been studied under conditions of high pressure (refer to Figure 2). Phase III is a period of partial or total early inhibition of the H2 reflex and lasts from 10-60 ms after the S1 stimulus; phase IV (the rebound) lasts from 60-250 ms and is a period of partial, normal, or supranormal responsiveness; phase V lasts from 300-5000 ms and is characterised by initially deep, then declining, late inhibition. Normal excitability is restored by some 5 s after the S1 stimulus. According to Hugon and colleagues (1983), the early part of phase III and the late phase V inhibition are caused by inhibitory discharges from recurrent Renshaw interneurones which fire after the S1 volley. Superimposed on this spinal inhibitory background are the afferent consequences of the H1 response - first a Ib burst and a Ia pause causing late phase III inhibition, followed by a Ib pause and a Ia burst causing the phase IV rebound - as the muscle first contracts and then relaxes.

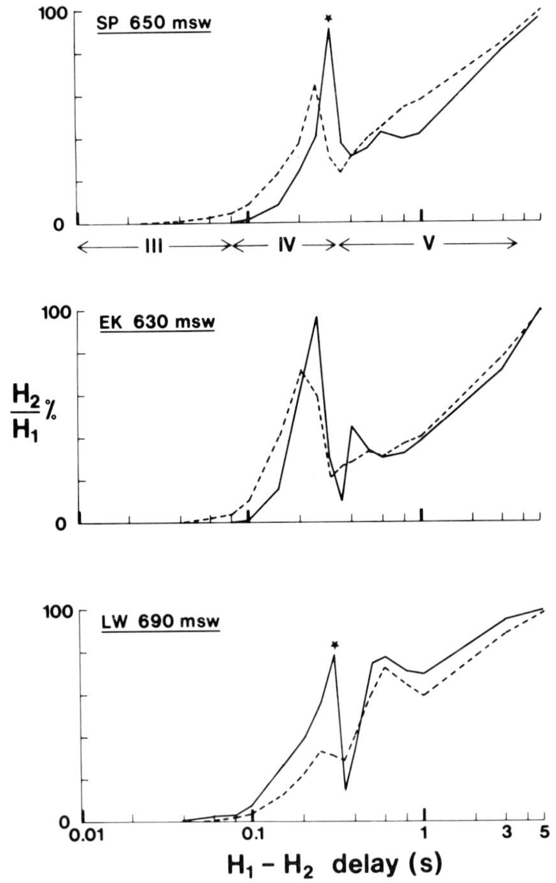

Fig. 2. H-reflex recovery cycles from three subjects during Atlantis III (10% trimix dive, 1981, Duke University). Dashed line shows average from 5-7 sessions during 8-week pre-dive period. Solid line is the result from 1 session at pressure, 10 measurements per delay setting. (, P<0.02). Paillard's phases III, IV and V (see text) are indicated below the top trace.*

Roll *et al.*(1978) produced recovery curves for one diver compressed to 610 msw in heliox; these were characterised by a depression of the entire curve during compression and saturation. In a shallower 420 msw saturation dive, Harris (1979b) observed a depressed phase IV in one subject and an elevated phase IV in the other, the two subjects then presenting quite different changes from each other for the rest of the dive. In seadragon IV, 300 msw also in heliox, H-reflex recovery cycles were consistently unchanged at depth (M.Hugon, unpublished data). In dives to 700 msw with the monkey, an enhanced and extended phase IV facilation was observed after fast compression (Hugon *et al.*, 1981). The dependence of the derived curve on the subject's posture, state of arousal, electrode placement etc make it critical that conditions are exactly equivalent on different recording days. These variable results may reflect experimental difficulties of this kind.

Recently, Harris and Bennett (1984a) obtained data from six divers all of whom had been compressed to 650 msw in 10% trimix (Atlantis II, III). Recovery curves from three subjects in Atlantis III are shown in Fig.2. The facilitated phase IV rebounds in these normalised curves is further evidence in favour of an increased discharge from the muscle spindle Ia afferents. However, the combined recovery cycle derived from the curves of 6 subjects at 650 msw revealed no consistent changes in early (III) or late (V) inhibitory phases : these data do not support recent tentative hypotheses regarding pressure effects on recurrent Renshaw activity (Hugon *et al.*, 1983). (However, we should not under-estimate the possible effects of 10% Nitrogen at these pressures in perhaps concealing an underlying pressure effect (Harris and Bennett, 1984a)). In Fig.2 it is also clear that the phase IV rebound is delayed; indeed, in 5 out of 6 subjects at 650 msw the phase IV peak was delayed by 25-50 ms. If the rebound is causally phase-locked to the muscle contraction, a delay in the rebound could only be due to a slowing of the twitch contraction at 650 msw.

IV. EFFECTS OF PRESSURE ON MUSCLE CONTRACTION DYNAMICS

Pressure-induced changes in the rate of muscle contraction were first suspected by Hugon, Seki and Matsuda (M.Hugon, unpublished data) in the 300 msw heliox dive, Seadragon IV; here, an H-reflex twitch in a tonically activated muscle took 10 ms longer than normal to reach peak tension. During Atlantis III the contraction time of H-reflexes and M-responses in the soleus was increased by 10-23% (12-27 ms) during compression (Harris and Bennett, 1982, 1983), while Parmentier *et al.*(1984) reported an average increase of 5-11% (6-13ms) in contraction time. The mechanism of this effect is unknown but since the latency of contraction remains unchanged, and there are no reported changes in action potential duration that correlate with the slowing of contraction (Harris and Bennett, 1983), it has to be assumed that it is an intramuscular phenomenon.

More than fifty years ago it was shown in isolated frog skeletal and heart muscle that pressure caused increased peak tension in twitch contraction (for a review see Macdonald, 1975). However, the measurement of maximum voluntary contraction in human divers at great pressure has usually yielded normal results (Gelfand and Lambertsen, 1978; Bolstad *et al.*, 1984), while the increased twitch responses observed at pressure were attributed entirely to increases in reflex loop excitability (Roll *et al.*, 1978; Harris, 1979a, b).

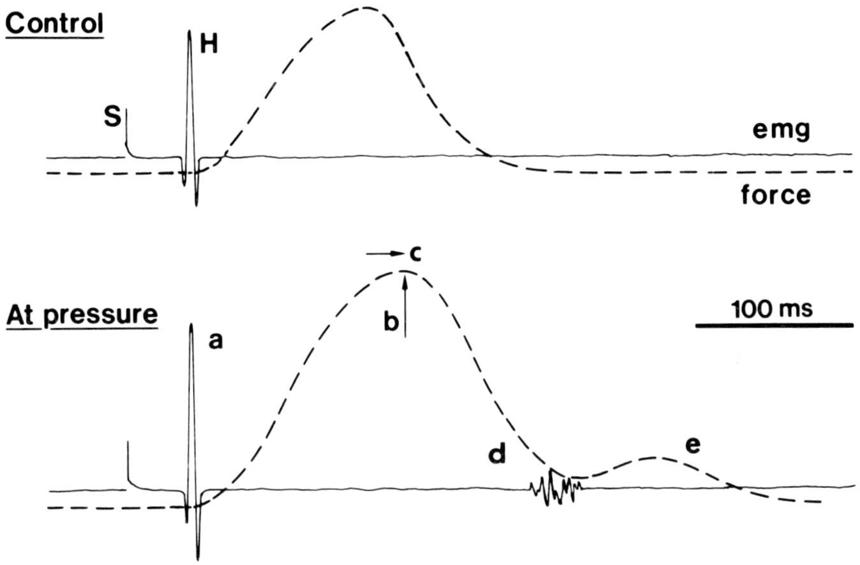

Fig.3. *Diagrams showing the main effects of pressure on EMG and force responses. Top control trace shows stimulus artefact (S), H-reflex action potential (H), EMG (solid line) and force trace (dashed line). Bottom trace shows that at pressure H-reflex responsiveness is increased (a), peak force of contraction is increased (b), the contraction may be slowed (c), afferent re-excitation (d)results in secondary or clonic contraction (e).*

It is now known that the tension developed by human anti-gravity muscles during twitch contractions at pressure can be 40%-70% higher than normal, an effect which is independent of, and additional to, any increase in response due to hyperreflexia (Harris and Bennett, 1983; Parmentier *et al.*, 1984). Unlike reflex excitability, peak force of contraction always returned to normal by the end of decompression. Furthermore, indirect electrical stimulation at 20 Hz of the adductor pollicis (McKenzie, 1977), and direct (percutaneous) 50 Hz tetanic stimulation of the quadriceps (Bolstad *et al.*, 1984) also resulted in increases of 15%-60% in the contraction force at pressure. It appears that compression can modify the way some critical substance or site is used during excitation-contraction coupling, or acts directly on the mechanochemistry of tension development (Harris and Bennett, 1983). A summary diagram of the reflex and intramuscular effects of pressure is shown in Fig.3.

Changes such as these which occur in muscle contraction dynamics at pressure can be expected to affect any muscle activity which consists of twitch-type contractions. The amplitude of isotonic contractions or the tension developed during isometric activity will be exaggerated solely by virtue of pressure effects on the muscle. The possible importance of this in the aetiology of tremor will be discussed in the final section.

V. LOSS OF REFLEX CONTROL AT PRESSURE : A MECHANISM FOR TREMOR ?

Recent work in the cat strongly suggests that high pressure tremor is mainly caused by segmental oscillatory mechanisms working upon a supraspinally facilitated background. Fagni *et al.*(1982) showed that the olivo-cerebellar pathway, which is essential for harmaline-induced 8-13 Hz tremor, is not required for the induction of pressure tremor : cats in which this pathway had been cut to abolish the harmaline tremor nevertheless developed an 8-13 Hz tremor upon compression. Even spinal cats were shown to develop a form of tremor upon compression and to be hyperreflexic at pressure in the limbs caudal to transection.

In man, some studies of clonic contractions at the ankle have lent weight to the segmental oscillatory concept. H-reflex stimulation at 5-second intervals has been used to modulate the EMG activity of the tonically-contracted soleus muscle, according to the methodology of Bouaziz *et al.*, 1975. The EMG records normally reveal a highly damped clonus of 2-3 beats duration following the isometric reflex twitch in the muscle (Bouaziz *et al.*, 1975; Harris and Bennett, 1984b). The relaxation phase of the twitch generates a large afferent Ia burst and Ib pause equivalent to the phase IV of the recovery cycle. This volley is apparently large enough to excite sufficient motor units to cause a second smaller contraction, which in turn gives rise to a second afferent burst during relaxation. The afferent bursts (rebounds) are separated by silent periods in the EMG corresponding to the inhibition of voluntary activity by a strong Ib burst and Ia pause during each contraction (Fig.4).

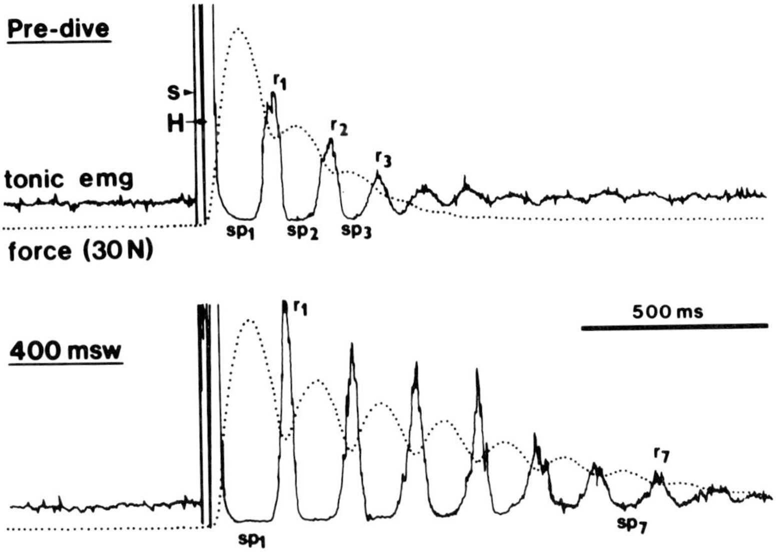

Fig.4. *Modulation of tonically activated soleus electromyogram(emg) by a single H-reflex during Atlantis IV (5% trimix, 1982, Duke University). Activation = 30 N force isometric plantarflexion. Rectified, averaged (n=100) EMG activity is interrupted by a silent period (sp1) which coincides with the rising phase of the H1 twitch. This is followed by a rebound (r1) of emg activity to a higher than voluntary level and a second contraction follows with a second silent period(sp2) and so on. Oscillations are usually damped out within 2-3 clonic beats; this subject was unusual, being "underdamped" before the dive. At pressure contraction was stronger and slowed, rebounds were larger and slowed, and damping was poor.*

At pressure the system appears underdamped with one or two extra beats per response, especially during hyperreflexia (Harris and Bennett, 1984b). In a recent 650 msw dive with 5% trimix (Atlantis IV) when T-reflexes were markedly increased, the isometric clonus was frank and sustained with the ankle oscillating sinusoidally for several seconds, as shown in Fig.4 (Parmentier *et al.*, 1984). When the muscle contraction was seen to slow, a correlated slowing of the clonic frequency occurred (Harris and Bennett, 1984b). Since the frequency of the "clonus" appears to be partly determined by the mechanical resonance of the musculo-skeletal elements in the limb, this motor control problem appears to be peripherally mediated and due to oscillations in the neuromuscular reflex loop.

Why does compression result in underdamping and loss of reflex control ? First, the muscle response to the efferent volley is greater than it should be (contraction force *per se* increased). This would result in larger than expected inhibitory (Ib) and excitatory (Ia, II) afferent bursts. Secondly, enhanced stretch receptor excitability further increases the afferent Ia volley : contractile and sensory factors combine to markedly increase reflex loop gain. It is not known whether this is sufficient to account for the evident level of afferent re-excitation of the alpha motoneurones, but the recent deep dives in trimix provided scant evidence that spinal recurrent inhibitory mechanisms are other than normal. However, since the H-reflex is slightly facilitated at depth, notwithstanding the doubt about what an H-reflex really is, this could arise from a central disinhibitory process of some kind, possibly of mid-brain origin, and possibly acting presynaptically (Delwaide and Delbecq, 1973; Roll *et al.*, 1978; Harris, 1979b).

When considering whether hyperbaric hyperreflexia can also be responsible for hyperbaric tremor, it must be borne in mind that there is no general agreement about the mechanisms of normal (physiological) and pathological tremors in normobaric man. Certain research groups have attributed physiological tremor solely to the properties of the monosynaptic reflex servo-loop (Lippold, 1970; Joyce and Rack, 1974; Matthews and Muir, 1980); others have proposed alternative mechanisms based on the grouped firing of motor units (Elble and Randall, 1976; Allum, *et al.*, 1978; Stiles, 1981). Several authors have proposed that reflex loop oscillations are one of a number of components all of which may be involved in tremor

(Elbe and Randall, 1978; Stiles, 1981; Brown *et al.*, 1982). Among these, Elble and Randall have shown convincingly that normal hand tremor has a neural component arising from grouped firing of random motoneurone spikes, a small cardioballistic component (<10%), and a mechanical servo-loop component.

From this review of the effects of pressure on reflex-loop physiology, we propose that the tremor of the HPNS is due to an exaggeration of the mechanical servo-loop component of normal "physiological" tremor. We agree with Fagni *et al.* (1982) that a cerebellar or centrally-driven mechanism of tremor is not needed to explain the tremor in the HPNS, and suggest that the key to understanding this motor problem is to recognize the disproportionately large afferent responses which result, under hyperbaric conditions, from a given efferent instruction to the limb muscle.

Acknowledgements. We gratefully acknowledge the contribution made by the divers and staff of the F G Hall Laboratory, Duke University Medical Center and of the AMTE Physiological Laboratory, Gosport who made it possible to carry out these projects. Funding support for the Atlantis dive series was provided by NIH grants HL07896 and GM24936, Shell Oil International, the US Navy Bureau of Medicine and Surgery, and the UK Ministry of Defence (PE). The views and opinions expressed in this article are the private ones of the authors.

REFERENCES

Allum, J.H.J., Dietz, V., Freund, H.J.(1978) Neuronal mechanisms underlying physiological tremor. J. Neurophysiol. 44, 557-571.

Bolstad, G., Gennser, M., Gustavson, M., Ornhagen, H.(1984) Pressure effects on human muscle. Proceedings of the IX Congress of the European Undersea Biomedical Society, in press.

Bonnet, M., Chouteau, J., Hugon, M., Imbert, G., Roll, J.P.(1973) Activités motrices et spontanées chez le singe sous heliox (99 ata). In : Hesser,CM.,Linnarsson, D.(eds), Swedish Journal Defence Medecine. First annual scientific meeting of the European Undersea Biomedical Society 9, 314-317.

Bouaziz, Z., Bouaziz, M., Hugon, M.(1975) Modulation of the soleus electromyogram during electrical stimulation of the medial gastrocnemius nerve in man. Electroencephalogr. Clin. Neurophysiol. 15, 31-42.

Brown, T.I.H., Rack, P.M.H.,Ross, H.F.(1982) Different types of tremor in the human thumb. J. Physiol. 332, 113-123.

Burke, D., Gandivia, S.C., Mckeon, B.(1983) The afferent volleys responsible for spinal proprioceptive feedback in man. J. Physiol. 339, 535-552.

Delwaide, P.J.(1976) Reflex physiology in man. In : Shahani, M.(ed), The motor system : neurophysiology and muscle mechanisms. Elsevier, Amsterdam, p.169.

Delwaide, P.J., Cordonnier, M., Charlier, M.(1976) Functional relationships between myotatic reflex arcs of the lower limb in man : investigations by excitability curves. J. Neurol. Neurosug. Psychiatr. 39, 545-554.

Delwaide, P.J., Delbecq, P.(1973) Vestibular influences on proprioceptive reflexes of the lower limb in normal man. In : Desmedt, J.E.(ed) New developments in electromyography and clinical neurophysiology, Vol. 3, Karger, Basel, p.336.

Desmedt, J.E. & al.(1973) A discussion of the methodology of the triceps surae T- and H-reflexes. In : Desmedt, J.E.(ed), New developments in electromyography and clinical neurophysiology, Vol.3, Karger, Basel, p.773.

Elble, R.J., Randall, J.E.(1976) Motor-unit activity responsible for 8 to 12 Hz component of human physiological finger tremor. J. Neurophysiol. 39, 370-383.

Elbe, R.J., Randall, J.E.(1978) Mechanistic components of normal hand tremor. Electroencephalogr. Clin. Neurophysiol. 44, 72-82.

Fagni, L., Weiss, M., Pellet, J., Hugon, M.(1982) The possible mechanisms of the high pressure-induced motor disturbances in the cat. Electroencephalogr. Clin. Neurophysiol. 53, 590-601.

Gelfand, R., Lambertsen, C.J.(1978) Muscle strength and coordination. In : Lambertsen, C.J., Gelfand, R., Clark, J.M.(eds), Predictive studies IV : Work capability and physiological effects in He-O2 excursions to pressures of 400-800-1200 and

1600 feet of sea water. Inst. Environ. Med., Univ. of Pennsylvania Medical center, Philadelphia, P.E5.1.

Harris, D.J. (1979a) Observations on the knee-jerk reflex in oxygen-helium at 31 and 43 bars. Undersea Biomed.Res. 6(1), 55-74.

Harris, D.J. (1979b). Hyperbaric hyperreflexia : tendon-jerk and Hoffmann reflexes in man 43 bar. Electroencephalogr. Clin. Neurophysiol. 47, 680-692.

Harris, D.J., Bennett, P.B. (1982) Muscle twitch contractions appear to be slower and stronger in man during deep saturation diving (Abstract). Undersea Biomed. Res. (Suppl.) 9, 22.

Harris, D.J., Bennett, P.B. (1983). Force and duration of muscle twitch contractions in humans at pressures up to 70 bar. J. Appl. Physiol.: Respirat. Exercice Physiol. 54, 1209-1215.

Harris, D.J., Bennett, P.B. (1984a) Soleus H-reflex studies in man at 40-70 bar in helium-oxygen-nitrogen (trimix). 11(1), 49-64.

Harris, D.J., Bennett, P.B. (1984b). Alteration of soleus electromyogram silent period in man at pressures up to 70 bar. In : Shilling, C.W. (ed), Underwater physiology VIII : Proceedings of the eighth symposium on underwater physiology. Undersea Medical Society Inc., Bethesda Md. In press.

Hugon, M. (1973) Proprioceptive reflexes and the H-reflex : methodology of the H-reflex in man. In : Desmedt, J.E. (ed), New Developments in electromyography and clinical neurophysiology, vol.3, Karger, Basel, p.227.

Hugon, M., Brauer, R.W. (1981) Concluding remarks on the mammalian nervous system at pressure. In : Obal, F., Benedek, G. (eds), Adv. Physiol. Sci., Vol.18, Environmental physiology. Academiai Kiado, Budapest, Pergamon, p.209.

Hugon, M., Fagni, L., Rostain, J.C., Seki, K. (1981) Somatic evoked potentials and reflexes in monkey during saturation dives in dry chamber. In : Bachrach, A.J., Matzen, M.M. (eds), Underwater physiology VII. Proceedings of the seventh symposium on underwater physiology. Undersea Medical Society, Bethesda Md. p.381.

Hugon, M., Fagni, L., Seki, K. (1983) Deep sea diving : human performance and motor control under hyperbaric conditions with inert gas. In : Desmedt, J.E., Advances in neurology, vol.39, Motor control mechanisms in health and disease. Raven Press, New York, p.829.

Joyce, G.C., Rack, P.M.H.(1974) The effects of load and force on tremor at the human elbow joint. J. Physiol. 240, 375-396.

Lippold, O.C.J.(1970) Oscillation in the stretch reflex arc and the origin of the rhythmical 8-12 c/s component of physiological tremor. J. Physiol. 206, 359-382.

Macdonald, A.G.(1975) Physiological aspects of deep sea biology. Cambridge University Press , Cambridge, p.159.

Matthews, P.B.C., Muir, R.B.(1980) Comparison of electromyogram spectra with force spectra during human elbow tremor. J. Physiol. 302, 427-441.

McKenzie, R.S.(1977) Changes in the indirect response of voluntary muscle in man during prolonged exposure to raised ambient pressure. J. Physiol. 272, 34-35p.

Paillard, J.(1955) Reflexes et régulations d'origine prorioceptive chez l'homme. Thèse Fac. des Sciences de Paris (série A n° 2858-3729). Arnette, Paris.

Parmentier, J.L., Harris, D.J., Bennett, P.B.(1984) Central and peripheral contributions to hyperbaric hyperreflexia. J. Appl. Physiol.: Respirat. Environ. Exercise Physiol. (in press).

Roll, J.P., Lacour, M., Hugon, M., Bonnet, M.(1978) Spinal reflex activity in man under hyperbaric heliox conditions (31 and 62 ata). In : Shilling, C.W., Kent, M.B.(eds), Underwater Physiology VI. Proceedings of the sixth symposium on underwater Physiology. Fed. Am. Soc. Exp. Biol., Bethesda Md.,p.21.

Stiles, R.N.(1981) Mechanical and neural feedback factors in postural hand tremor of normal subjects. J. Neurophysiol. 44, 40-59.

SUBJECT INDEX

A

B

C

D

E

F

K

L

M

N

R

S

T

V

W

X